W9-CCF-232

Clinic on Library Applications of Data Processing: 1988

Design and Evaluation of Computer/Human Interfaces:

Issues for Librarians and Information Scientists

Edited by
MARTIN A. SIEGEL

Graduate School of Library and Information Science
University of Illinois at Urbana-Champaign

ISBN 0-87845-072-6 ISSN 0069-4789

CONTENTS

Introduction

A computer/human interface functions at the intersection of machines, people, information, and tasks. Growing emphasis within the library and information science community has been placed on the development of these interfaces—an aspect of design loosely termed user friendliness. This emphasis becomes increasingly important as the number of computers in libraries continues to rise along with their use by nonspecialists.

This clinic focused on the interface requirements of two main user groups: the professionals and staff behind the desk and the patrons in front of the desk. The clinic sessions were organized around a series of presentations to help the professional examine the issues surrounding the design, selection, and evaluation of computer/human interfaces found in libraries. Anyone who has used a word processor, struggled through the intricacies of online database searches, or wondered about improvements to online catalog systems will benefit from reading the papers in this volume. These papers emphasize people rather than technology.

The year 1988 marked the 25th Annual Clinic on Library Applications of Data Processing. Therefore, it was appropriate for three members of the Graduate School of Library and Information Science faculty to provide a quarter-century retrospective focusing on three major clinic themes: public services, technical services, and administrative/management concerns. This volume begins with these preconference papers.

MARTIN A. SIEGEL
Editor

LINDA C. SMITH

Associate Professor
Graduate School of Library and Information Science
University of Illinois at Urbana-Champaign

From Data Processing to Knowledge Engineering: The Impact of Automation on Public Services

INTRODUCTION

The title of this paper is a reminder that, although the name of the Clinic has remained the same for the past 25 years, the goals and means for the application of automation to public services have evolved. Although the term *knowledge engineering* is presently used to denote the process of building an expert system (Waterman, 1986, p. 5), dictionary definitions of the component words suggest a wider possible scope. While *data* can be defined as "a group of facts or statistics" (Webster's Dictionary, 1956, p. 210) and *to process* is defined as "to put through the steps of a prescribed procedure" (American Heritage Dictionary, 1970, p. 561) or "to handle in a routine, orderly manner" (Random House Dictionary, 1980, p. 713), *knowledge engineering* covers a wider domain. *Knowledge* is defined variously as "familiarity, awareness, or understanding gained through experience or study" (American Heritage Dictionary, p. 393); or "that which is known; the sum or range of what has been perceived, discovered or inferred" (p. 393). Finally, *engineering* may be thought of as "the application of scientific principles to practical ends as the design, construction, and operation of efficient and economic structures, equipment, and systems" (p. 239). This review of the history of automation in public services, as documented in papers presented at past data processing clinics, reflects the change in emphasis over time from data to knowledge and from processing to engineering. This paper has five parts: a brief chronology of the clinics, an analysis of various public services and the effects of automation thereon, the roles of library staff, the effects on users, and future prospects.

CHRONOLOGY

The first Clinic on Library Applications of Data Processing was held at the Illini Union on the Urbana-Champaign campus of the University of Illinois, April 28 - May 1, 1963 under the sponsorship of the University of Illinois Graduate School of Library Science. Writing in the Foreword to the Clinic proceedings, Herbert Goldhor (1964) provides the rationale for sponsoring such a Clinic:

> Starting from the proposition that it is the proper function of a library school to give leadership to the profession, the Faculty of the University of Illinois Graduate School of Library Science have from time to time attempted to identify major challenge problems of our age and to formulate appropriate responses to them. (p. iii)

The Clinic was one part of the School's response considering the role of data processing in meeting the challenges facing libraries. The first seven Clinics, during 1963-1969, dealt with data processing in relation to all major aspects of library operations. Three different individuals served as editors of those early volumes: Herbert Goldhor (1963-64, 1966), Frances B. Jenkins (1965), and Dewey E. Carroll (1967-69).

Beginning in 1970, each Clinic had a theme concentrating on a specific aspect of library data processing. Although various aspects of the automation of public services were touched on in several of the Clinics, only two had this topic as the main focus: the 1975 Clinic entitled *The Use of Computers in Literature Searching and Related Reference Activities in Libraries* (Lancaster, 1976) and the 1987 Clinic on *Questions and Answers: Strategies for Using the Electronic Reference Collection* (Smith, 1989).

Discussion of public services automation in the 1960's Clinics was confined primarily to circulation systems, particularly in large academic libraries, and to information retrieval systems in use in special libraries and information centers. The dominance of technical services applications during this period is not surprising. John A. Wertz (1965) advised Clinic participants to "search the library for operations which consist only of the clerical tasks of rearranging the format of information, the simple comparison of one datum to another, or the creation of ordered lists of data" (p. 114), for these were the things a computer could do. He forecast that the impact of automation would be greatest upon the technical service and administrative areas, "partly because the computer cannot change the intellectual process of questioning (and decoding the question) and partly because the technical processes are most open for improvement" (p. 115). I. A. Warheit (1972) noted that it was not surprising that the Clinic included no reports of online, interactive reference systems because machine-readable files of large, general collections had not yet been built (p. 18). By 1975 there was enough activity to warrant devoting a Clinic to computer applications in reference activities (Lancaster, 1976), and at the 1981 Clinic, Ronald L. Wigington (1982) observed that "attention is now turning to improving the service-providing aspects of libraries, such as subject access and document delivery" (p. 5).

SERVICES

For the purposes of this paper, *public services* will include circulation, information retrieval/literature searching (both retrospective searching and current awareness/selective dissemination of information), interlibrary loan/document delivery, referral/community information services, question answering/ready reference, and information management. Rather than providing a strict chronology of developments as reported in the Clinics, each of these service areas is discussed.

Circulation

From 1963-74, Clinic proceedings contain many descriptions of circulation systems, demonstrating the evolution in design and capabilities as the underlying technology changed from punched card equipment to second generation computers, then to third generation computers with online access and, finally, to minicomputers. Early implementations had to deal with severe limitations in the technology. For example, Benjamin Courtright (1966) of Johns Hopkins University reported that "the IBM 1401 in use has 8000 positions of core storage, four 729 magnetic tape drives (no disk storage), and a fairly complete array of special features on the central processing unit. The 8K storage limitation is a decided nuisance, but so far it has been possible, if somewhat awkward, to program around it. The daily circulation control update program, for example, has nine overlays" (p. 27). Because of the difficulty of programming, there were often delays in implementation. Ralph H. Parker (1964) of the University of Missouri noted that their circulation system was already nearly two years behind the original schedule and commented that "pioneering is a slow business" (p. 54).

The first report on an online circulation system came when Robert A. Kennedy (1970) described Bell Laboratories' BELLREL (Bell Laboratories Real-Time Loan) system. In its initial configuration, the system linked two terminals in each of the company's three largest libraries to an IBM 360-40 computer at Murray Hill, New Jersey. The system was designed to process loans, returns, reservations, and a range of information queries in real time; in addition, batch operations provided multiple reports to aid in control and management of library resources. Hugh C. Atkinson's (1972) report on the Ohio State online circulation system noted that "the system should be one which would speak to the problems of its users rather than simply the problems of the library" (p. 23). A major problem for users prior to the introduction of the online circulation system was the location of materials on a large campus with a decentralized library system. He reported that circulation rose over 40 percent in the first eighteen months of system operation, reflecting the increased ease with which users could learn of the availability of wanted materials. In the same year, Joseph T. Paulukonis (1972) described the online real-time self-service circulation system at Northwestern University which allowed users to charge materials themselves.

The 1974 Clinic on the applications of minicomputers introduced the concept of turnkey systems with Dennis N. Beaumont's (1974) description of the LIBS 100 system developed by Computer Library Services, Inc. (CLSI). He outlined the advantages of this approach to automation:

> This system is delivered ready to use, including the equipment and programming necessary to meet the library's unique requirements. The library need not hire or retain specialists in library data processing applications. Libraries may now automate without major investment in system development or specially trained staff, and without start-up difficulties frequently associated with this process. (p. 55-56)

At the same Clinic, Wayne Davison (1974) of Stanford pointed out that minicomputers are suitable technology for circulation control systems because of their reliability and availability. Libraries could afford to have dedicated minicomputers supporting their circulation activities.

In the past few years, discussion of circulation systems at the Clinics has been confined to consideration of online public inquiry to check library holdings and circulation status. Margaret Beckman (1981) described the features of the online circulation system at the University of Guelph. The inquiry module of the circulation system was designed with self-instructing display screens to lead from one command, instruction, or question to another. Beckman reported that in fall 1979 a remote access capability was implemented, allowing any terminal on the campus network to access the circulation system.

Information Retrieval/Literature Searching

Online access to circulation systems emphasizes know item searching. Since Donald H. Kraft (1964) reported on the application of IBM equipment to Keyword-In-Context (KWIC) indexing and the Selective Dissemination of Information (SDI) at the first Clinic, the use of computers as aids in subject searching has become much more widespread. Presentations at the Clinic first described batch searching systems and then noted the growing availability of online access to machine-readable databases. More recently, the emphasis has turned to end user searching of remote databases, online catalogs, and databases on CD-ROM.

Kraft characterized SDI as a new role for the library, that of an active disseminator of information. His description of SDI notes the potential advantages without acknowledging possible pitfalls:

> SDI is like an electronic traffic director of information, analyzing and routing it to those who have a need to know. This is a tool with which one can match or compare his particular interests against the flood of paper. Through the use of this technique, one automatically receives only the specific information that he really wants. SDI provides the librarian with the means to extend himself almost infinitely in his ability to service his users' information needs promptly. (p. 143)

Richard W. Counts (1967) of the Aerospace Research Applications Center at Indiana University described the elaborate computer system supporting their current awareness service. Because interest profile development and maintenance was such a labor-intensive activity, the Center planned to undertake research projects to investigate the feasibility of automated profile maintenance.

Provision of current awareness service was not without problems. G. E. Randall (1973) of the IBM Research Center Library described a centralized computer-based current awareness service with a database including all IBM internal reports, IBM patent disclosures, IBM operating and systems manuals, the National Technical Information Service (NTIS) tapes for external reports, and COMPENDEX (Computerized Engineering Index) for the open literature. Because the database included citations to some material not in the IBM collections, not all requests could be satisfied. Randall commented on the resulting stress as follows: "Give a library a fixed budget, engender requests which cannot be supplied from the collection supported by that budget and you promote ulcers in the responsible librarian" (p. 120).

SDI services were provided by special libraries and this was initially true for retrospective searching services as well. Response time for the original batch-searching systems which accepted requests from other institutions could be measured in weeks rather than seconds. Louise Darling (1966) described the MEDLARS search service at UCLA. She explained that processing a batch of 10 to 15 searches through the entire MEDLARS file could take from four and a half to five hours. She observed that once the search service was "operating routinely and well, requests should be processed in two weeks, with emergency service in one, providing machine time is available and search volume within reason" (p. 99). At the same Clinic, M. M. Kessler (1966) of MIT offered a view of things to come with his description of the Technical Information Project which made use of a time-sharing system for literature searching. Kessler felt it necessary to define *on line* for his listeners as the capability that "a dialogue can take place between the person and the machine" (p. 9).

By 1975, online searching was well established in special libraries and was increasingly available in academic libraries. Online systems had all of the capabilities of offline, batch-processing systems without any of their major disadvantages. They supported interactive searching and provided results with very little time delay. Roger K. Summit and Sally J. Drew (1976) reported preliminary findings of the first major study of online searching in public libraries. Their conclusions were optimistic:

> The first year of the project has shown that computerized search allows the public library to offer in-depth search service in diverse intellectual fields, and in fields in which the reference librarian is not expert. Computerized reference retrieval has been of great use to the public libraries that do not have a large reference collection. It also can be more cost effective than manual search for many topic areas even when a large reference collection is available. The public has shown great interest in the service and has expressed very positive evaluations of the results obtained. We anxiously await the second year, which will demonstrate whether or not the public is willing to pay for such services. (p. 95)

Throughout the 1970s, library users had to rely on librarians acting as intermediaries to search the available online systems. This was a situation causing concern to some of the Clinic speakers. Davis McCarn (1973) of the National Library of Medicine noted that online had not yet reached the user because online systems were not being used directly by those with the information need. Indeed, evaluations of MEDLINE completed in 1972 and 1973 showed that over 75 percent of searches were run in the absence of the requester (pp. 3-4). Sally Bachelder (1976) described the New York Times Information Bank, explaining that its use of conversational English and avoidance of function keys made the system simple to master. Nonetheless, most subscribers restricted use to librarians to hold down costs. She concluded that "while today's technology has made the Information Bank and its user orientation possible, today's economy is preventing the user from gaining full control of the inquiry process" (p. 22).

Widespread end user access to online systems had to await new developments in the technology. The possibility was foreseen by A. E. Negus (1974) when he suggested that minicomputers connected to a network could provide a simplified and standard dialogue which would make the various systems to be accessed look the same to the user, obviating the need for separate training in the use of each different system (p. 100). David M. Wax (1976) described the need for an even more elaborate form of machine assistance to users:

> the development of a compiler that would not only make the systems compatible, but would also relate the user's information needs to the appropriate data bases and translate the search question into appropriate strategies with appropriate vocabularies to search these data bases on whichever systems they happen to be accessible. (p. 80-81)

Martha E. Williams (1980) outlined in some detail the transparency aids that could be the components of a transparent information system designed to support searching by a wide variety of types of users. Transparency aids would include converters, selectors, evaluators, analyzers, and routers. Results of the Individualized Instruction for Data Access (IIDA) project as reported by Thomas T. Hewett and Charles T. Meadow (1981) suggested that software aids could allow end users to do searches comparable in utility to those done by a professional searcher. By the mid-1980s, minicomputer software products such as the Sci-Mate Searcher described by David E. Toliver (1987) of the Institute for Scientific Information could minimize differences among access protocols, retrieval commands/responses, and database structures, thus greatly simplifying searching for the end user. An alternative to buying interface software is to develop it locally, as described by William H. Mischo and Melvin G. DeSart (1989) of the University of Illinois. They explain that such customized microcomputer interface software can allow end user searching of external databases as an option within an online catalog interface. This is particularly advantageous in science libraries needing enhanced online access to contents of periodicals.

As an alternative to incurring charges for end user searching of external databases, libraries can now purchase selected databases on CD-ROM. The 1987 Clinic was the first to explore the possibilities of this technology. Terry Noreault (1989) of OCLC noted that CD-ROM is best for fairly large databases with many potential subscribers. Likely applications for CD-ROM include supplementing online searching, providing access to reference material, and storing some material in full text. Online systems will still be used when the databases are large, currency of information is important, or the database is not used often enough to warrant purchase in CD-ROM format. Karen Havill Bingham and Barton M. Clark (1989) reported the University of Illinois at Urbana-Champaign's experience with InfoTrac. They noted that student response was enthusiastic, but that students are frustrated by the limitation of one database per workstation. In particular, students would like to be able to link InfoTrac to the library's online catalog.

It has taken much longer than anticipated to couple catalogs and information retrieval systems. Frederick G. Kilgour (1969) spoke on the initial system design for the Ohio College Library Center. He predicted:

> Soon after the shared-cataloging project goes into operation, there will be activated a bibliographic information retrieval system which will allow users of Ohio libraries to obtain rapid and complete searches under subjects. An effort will be made to increase amounts of subject indexing, which are presently inadequate in all libraries. (p. 86)

Twenty years later, Charles R. Hildreth (1989) was finally able to report a convergence of online catalogs and conventional information retrieval systems. He suggested that the promise was clear: online catalogs could be better library catalogs than card catalogs, as well as powerful and usable interactive retrieval systems. At the same conference, Betsy Baker (1989) reported that networks were increasingly moving in new directions to support more public service needs. She cited as evidence OCLC's cooperative work with BRS to provide subject access to a cross-disciplinary database of OCLC records available using BRS's searching software. In addition, OCLC was providing subject-oriented subsets of the OCLC database on CD-ROM, gateway searching access to BRS databases from the OCLC terminal, and improved dial-access capability.

Interlibrary Loan/Document Delivery

With the advent of online literature searching, library users were faced with the dilemma that references to relevant literature could be found in minutes, but the actual documents could take days or weeks to track down. In addition, because it was so easy to obtain references to documents not held locally, demand for interlibrary loan services grew. A natural response was to see if technology could be enlisted to speed up the interlibrary loan process. Joseph Becker (1973) reported that ARL was considering using a computer as an "electronic mailbox" to store interlibrary loan request messages for particular institutions and transmit them automatically according to predetermined schedules or on demand. He outlined the possible functions of such a system in more detail:

> A national system of interlibrary communication for interlibrary loan

> would not only route messages more effectively, but it could also utilize companion computer programs to manage and administer the operation of the total system. Thus a computer could handle billing, maintain the statistics, do accounting, keep track of copyright royalties, etc. In time, with heuristic programming and a directory of holdings, a computer might even learn to switch incoming requests automatically to those institutions in the network that have the highest response potential. (p. 176)

Dennis Oliver (1985), in his discussion of electronic mail services in the library, reported that bibliographic utilities had moved closer to electronic messaging through the implementation of interlibrary loan subsystems.

Irwin H. Pizer's (1972) paper on online technology in a library network serves as a reminder that technology alone is not enough. He noted that an example of the possible power of an online network involves the interlibrary loan process. The State University of New York (SUNY) network performed all of the necessary programming and testing of procedures to support interlibrary loan, and the system was declared operable in 1969. He reported that "as the librarians of the member institutions were faced with the reality of accepting a larger number of interlibrary loan requests than they had been accustomed to receiving, they balked. The Network Advisory Council felt that the anticipated avalanche of requests would render normal service in this area unworkable, thus the automatic interlibrary loan procedure was never tried, even on a limited basis" (p. 58).

Referral/Community Information Services

Just as interlibrary loan allows a library to facilitate user access to printed sources of information not held locally, referral or community information services can link library users to organizations or individuals relevant to their needs and interests. Technology has been used to support and enhance this service as well. Perhaps the best example of this concept is provided in MAGGIE'S PLACE at the Pikes Peak Library District, as described by Ken Dowlin (1987). Files maintained online for consultation by the public included listings of community agencies, clubs, and organizations; adult education courses; an events calendar; and daycare centers. The online system also supported a community-wide public carpool system and a transit information system to provide schedules of the city bus system online.

Question Answering/Ready Reference

MAGGIE'S PLACE provides one example of online support of question answering or ready reference. Increasingly, librarians have recognized that other tools, including the online catalog and online information retrieval systems, are also valuable aids to ready reference. In fact, Hugh Atkinson (1972) reported that, although the online circulation system had not been designed with the needs of reference librarians in mind, it was proving to be an important part of reference services. The terminal provided in the reference department was used for answering catalog information questions, matching bibliographic sources against library holdings. This example illustrates an observation made by historian of technology Derek de Solla Price (1980) at that year's Clinic: "A new technology never just replaces the old method—it enables quite different styles of life to come into being. Furthermore, it is the very indirect results of a technology that are its most interesting and sometimes its most significant consequences" (p. 14).

Another precursor of widespread use of the terminal at the reference desk was the New York Times Information Bank. Sally Bachelder (1976) reported that subscribers made heavy use of the database for book and theater reviews, analyses of topics of current interest, information on state and federal legislators, and information on the economy (p. 20). The system was used when it was the only source that could supply the information, or when the time required for a manual search would be too great.

By 1987 the electronic reference collection had considerably expanded in scope. Charles R. Anderson (1989) described online ready reference in public libraries, equating ready reference with being able to find quick facts. He suggested that three elements are required for online ready reference: the terminal must be at the reference desk allowing easy connection to online vendors; someone capable of searching online must be scheduled at the reference desk at all times the library is open; and the library should not charge its patrons for responses to ready reference questions relying on database access. Virgil P. Diodato (1989) reported on online ready reference in academic libraries, and proposed that questions to be answered online require short periods of online time, small amounts of online costs, and brief retrieval of information. Examples include bibliographic verification, current information, information that requires coordination of several search concepts or terms, information for which no printed sources exist in the library, and definitions and directory information. At the same conference, Beth S. Woodard (1989) gave examples of how the online catalog can respond to many of the same types of ready reference questions.

Information Management

One other service is assisting researchers in management of their personal files. F. W. Lancaster (1976) suggested that assistance in the building and maintenance of personal files might be a needed service in universities, research institutes, industry, and government agencies (p. 153). Although Lancaster envisioned such files as being online, microcomputer software packages now offer an alternative. While this topic did not receive attention in subsequent Clinics, many academic and special librarians now do see this as a possible service, as described, for example, by Camille Wanat (1985) and Elizabeth H. Dow (1987).

ROLES OF STAFF

Having reviewed the growth and change of services as a result of applications of technology, it is appropriate to consider next the staff roles and responsibilities in providing those services. In her paper on competencies required of public services librarians to use new technologies, Danuta Nitecki (1983) argued that the basic competencies required of public services librarians to perform their primary functions today are the same whether or not automated resources are used. She enumerated and discussed five basic types of competencies, including the abilities to communicate with others, to analyze needs, to retrieve data, to instruct users, and to manage operations and supervise staff who provide services. There are new techniques to be mastered and new resources to be understood, but this was true of the print reference world as well. As one reviews discussions of staff roles in the various Clinic proceedings, two themes emerge: an attempt to distinguish intellectual from mechanical activities and an attempt to define the librarian's relationship to users.

Mechanical vs. Intellectual

Some early Clinic speakers felt the need to remind their audiences that not everything would be mechanized. Concluding her presentation on mechanization of routines in the IBM Advanced Systems Development Library, Marjorie Griffin (1964) noted that "since the library staff will be freed from these clerical tasks, they will be able to render more intellectual service; for no matter how much we accomplish in library automation, there is always need for intrinsic intellectual effort" (p. 95). In the same year, Seymour I. Taine (1964) described bibliographic data processing at the National Library of Medicine as MEDLARS was becoming operational.

> Certain tasks, such as indexing, cataloging, preparation of search requests, and proofreading will be performed by people while machines will perform such operations as storage, retrieval, and composition of bibliographic citations.

> In the development of a computer information retrieval system such as MEDLARS, there is a natural tendency to stress the data processing aspects at the expense of these humanly performed functions. Since it is indisputable that the mechanized portion of the system requires a tremendous effort to engineer, it becomes necessary to exercise care and discipline to avoid relegating the other parts to a kind of second-class status. This could be fatal. Difficult and complex as the data processing problems that confront us are, I have the conviction that the success or failure of MEDLARS will be more directly related to the non-mechanized elements. (p.124)

As online systems became available, there was an even greater interest in understanding the intellectual processes of the human and the mechanical processes of the machine combined through interactive processing. I. A. Warheit (1972) asserted that "the exploitation of the interplay between what the machine can do and what the human can do is what will really advance the state of library technology" (p. 19). Fifteen years later, in 1987, the issue rose again as developers sought to build expert systems for online searching and other aspects of reference work. As Stephen P. Harter (1989) pointed out, before this can be accomplished, system designers will have to know much more about how librarians search than is known at present.

Relationship to Users

Given that expert systems are quite unlikely to take over all situations in which public services librarians work with users, it is of interest to explore the impact of automation on librarians' relationships to users. In 1975, D. M. Wax (1976) noted that, while special librarians had long been involved in carrying out personalized bibliographic searches for library users, the role of the online searcher was a totally new one for the academic reference librarian (p. 81). He also observed that introduction of computer-based reference services required the librarian to adopt a more active role in user education and service promotion.

Others have noted the change in user expectations associated with computer-based reference services. In 1975, R. K. Summit and S. J. Drew (1976) commented that patrons using the online search service could require more of the reference librarian's time and were often more critical of the results than the typical reference patron (p. 95). Danuta Nitecki (1983) elaborated on this point:

> There seems to be an implied higher expectation of librarians to provide precise, accurate information with the use of online systems than was previously experienced. The costs—calculated with each instance of use of automated resources—and the visibility of errors—such as incorrect logic, improper selection of files or terms, and

> misspelling—both contribute to a rising sense of accountability among public services librarians using online sources.

She argued that the experiences of providing online services and the resulting sensitivity to efficiency, user satisfaction, and cost effectiveness should also contribute to development of library service philosophies in other aspects of public services.

As end user searching becomes more common, there is still a need for librarians to serve as consultants. In addition, reference librarians themselves may need to consult each other as one means of answering difficult questions. R. Bruce Briggs's (1976) report on a study of the user interface for bibliographic search services identified the need for remote access to system specialists in a network environment. Murray Turoff and Starr Roxanne Hiltz (1980) advocated the use of computer conferencing systems to support referential consulting networks, individuals willing to share their knowledge to answer questions. Richard T. Sweeney (1983) suggested that "the concept of librarians providing assistance to patrons, some of whom are in a distant location, while 'walking' them through a search is a promising continuation of the best part of our profession" (p. 60). Sheila Creth (1989) elaborated more fully on the possibilities of online reference service. The online question can be forwarded to the librarian with the subject expertise who will be able to provide the most knowledgeable assistance rather than having a question answered by the person who happened to be scheduled at the reference desk. In a fully automated, network environment the librarian responding to a reference question will also be more likely to use telecommunications to communicate with colleagues within the institution and nationally in responding to a particularly complex reference question.

EFFECTS ON USERS

In turning to a discussion of the effect of automation on library users, it is appropriate to recall Robert A. Kennedy's (1970) observation on his library's automated circulation system: "What BELLREL has meant to the individual library user is less well known. Many scientists and engineers have expressed strong technical interest. Many understand how the system works for them. Others, it seems, would accept conveyor belts or carrier pigeons, as long as information needs were met" (p. 30). Thus one must not let fascination with the technology obscure the ultimate aim of better satisfying users' information needs.

User Needs

Burton W. Adkinson (1964) recognized that data on information needs and uses could be gathered as a byproduct of mechanized systems. He cautioned that "it will take a sizable effort to be able to gather the right kind of such information and to make good use of it. Yet the effort needs to be made" (p. 7). M. M. Kessler (1966) reported that the early online retrieval system at MIT included a monitor system "so that everybody who is using the system gets recorded as who he is, how long he is using it, what questions he asks and so on" (p. 16). Another approach was to study users of existing systems to suggest requirements for future systems. In Ben-Ami Lipetz's (1970) report on an extensive study of users of the Yale University Library catalog, he explained: "Our study attempts to find out what our users want from a catalog, but it does not stop there. It also attempts to find out the extent to which our present card catalog satisfies the needs of the users. And, furthermore, it attempts to find out whether there are practical methods, manual or mechanized, to satisfy needs that are not now being met" (p. 44). He assessed the value of such research as follows: "With all the effort that has been going into research and development work on how to computerize catalogs, it would be nice to have more guidance on how to do it right" (p. 48).

User Satisfaction

Closely related to the concept of user needs is that of user satisfaction. It is appropriate to begin with Mooers's Law, cited by F. W. Lancaster (1976): "An information retrieval system will tend not to be used whenever it is more painful and troublesome for a customer to have information than for him not to have it" (p. 145). Lancaster suggested that convenience appears to be the single most important factor determining whether or not an information service will be used. This observation was echoed by Irwin H. Pizer (1972) in his discussion of online system design. He commented that "we learned very early that the threshold of user impatience when using an on-line system is very low. One can choose to ignore the problem, but one then finds that the system is regarded as unsatisfactory by many, and ignored altogether by others. Driving away the potential user is not the solution to the problem." (p. 65). Considering the convenience of remote access, Frederick G. Kilgour (1977) remarked that "not having to go to a library is a very important improvement in providing library service" (p. 8).

In a discussion of administrative considerations for library automation, Richard E. Chapin (1967) noted that the best-laid plans of any automated system can be hampered by the users. He advised that "they do not have to love the computer, but they should accept it. In order to meet this criterion, a system should be designed so as to provide the user with more information or better service. If possible, the user should be informed as to why automation is being installed and how it will work" (p. 64). Twenty years later, in his discussion of optical storage media, Terry Noreault (1989) suggested that these media would have a much larger effect on the delivery of reference service than simply through their direct application. He predicted that users would become more aware of the benefits of electronic information sources and expect to have access to them for all their information needs. Thus CD-ROM, while not appropriate for every information need, would educate end users about what is possible and thereby increase the demand for all forms of electronic information sources.

FUTURE PROSPECTS

While it may seem somewhat anomalous to conclude a historical review paper with a section on future prospects, the Clinic proceedings in fact include a number of forecasts regarding the future of automation in public services which are as yet only partially fulfilled. It is of interest to review these forecasts to suggest what remains to be done. The forecasts include prospects for remote access, integrated access, and future roles for librarians.

Remote Access

Forecasts on remote access are the first area to be examined. Joseph Becker (1966) found prospects for a national library network exciting:

> A library network is rich with the promise of a wholly new approach to the problem of gathering and retrieving essential information. It will transform libraries into active, rather than passive, sources of knowledge by permitting information exchange to flow in either direction between library and patron. The ability to confer with a library without necessarily visiting it, and the added ability to transfer selected information from library to home or office for individual use, should have a profound effect on the processes of research and education. (p. 5)

William J. Kubitz (1980) was able to describe in some detail the technology that would support remote access:

> The availability of low-cost computing and storage will make computers available and economical for everyone, whether in business, industry or the home. What we now call "microcomputers" will

> become as powerful as present-day large computers, but will sell for under $1000. Large central data banks will be formed as repositories of information. High-speed digital communications links will be readily available by way of satellite transmission in space and optical fibers on the ground. Digital communications will be brought into the home via the telephone system or cable TV or both. This will allow the user to call the central data banks using the home computer system. The home computer systems will have color TV, voice output, limited voice input, possibly a facsimile printer, and an associated flat panel character (book) display. (p.160)

Integrated Access

Remote access is a necessary but not sufficient condition to insure exploitation of the growing number of information resources in machine-readable form. Various Clinic speakers have outlined plans for integration of electronic resources to insure their coordinated use. Some of the proposals presuppose directories of information resources to support automatic switching of requests to appropriate resources. Thus Don Swanson (1967) envisioned catalogs and indexes that could guide one to the location of any item of material in the universe of recorded knowledge. He suggested that "from a single point of interrogation we should be guided either to *Chemical Abstracts,* to the National Union Catalog, or to a small catalog of special materials one of which, with further dialogue, would yield the answer" (p. 4). Glyn T. Evans (1973) provided examples of necessary linkages among files. The user finding serials citations while searching a database such as ERIC should then be led to a serials data file to impose bibliographic consistency. From there, the citations are passed to a Union List of Serials database to discern locations. From there, they go to a circulation file to assess availability and then back to the librarian or user, some data perhaps being forwarded to an interlibrary loan module either directly or following user decision. Evans further suggested that this should be automatic, with results of all searches presented to the user as a complete report in response to the search request. Charles R. Hildreth (1989) presented a contemporary view of the one-stop, self-service, information access and delivery station (or the "scholar's workstation"). Databases to which integrated access could be provided include locally created and maintained files, remotely published files stored and accessed locally, and external bibliographic and information files.

Charles T. Meadow (1989), also a speaker at the 1987 Clinic, offered an alternative vision, emphasizing different programs for different situations. He pointed out that there is not a universal reference book. He proposed that individual authors who can visualize a problem situation should design for that situation. A "situation" is a combination of a set of available information and potential users. For example, the physician interested in current information about drugs on the market has a different need than the research pharmacologist. Different groups may have not only different needs, but different searching skills as well.

New Roles

Meadow's vision of tailoring system design to users suggests new roles for librarians. Wigington (1982) predicted that "the information professional will fill a prime role in the design of man-machine dialogue and algorithms for use by persons less experienced in information system intricacies" (p. 11). Richard T. Sweeney (1983) predicted that librarians will be involved in building knowledge bases, "organizing the electronic information delivery and access, providing a quality filtering and synthesis process that reduces much of the redundant and irrelevant information to just what a practitioner or scholar needs" (p. 66).

In moving from data processing to knowledge engineering, much basic research needs to be done and this, too, has been acknowledged over the years by various Clinic speakers. Burton Adkinson (1964) noted the need for investigations into the nature of language and how it represents and conveys information. Joseph Becker (1966) saw the need for study of methods for automatically extracting meaning from a text, involving correlation of facts and inference of subject relationships from the complete content of articles and books. He rightly observed that this area would "require additional research before we can see clearly how such a capability will affect the duties of the reference librarian" (p. 3). David L. Waltz (1976), describing natural language question answering systems, predicted that the descendants of the systems he described "could eventually revolutionize the entire structure of libraries, as well as the lives of all those who use and benefit from libraries" (p. 144). Finally, Tamas E. Doszkocs (1987) identified several relevant research areas in addition to natural language processing techniques: expert systems, intelligent information management, and artificial intelligence knowledge representation techniques.

Although the Clinics have been sponsored from the beginning by a library school, relatively few papers specifically addressed the implications of library automation for library education. An exception was Don R. Swanson's (1967) paper, in which he suggested two reasons why automation might be expected to have an important impact on librarianship. First, questions on purposes and goals are raised that no one before thought of asking. This is reminiscent of the conclusion of Jesse Shera's (1964) essay on "Automation and the Reference Librarian" which stated that "a new understanding of librarianship may eventually prove to be the greatest single gift of automation to the library world" (p. 7).

A second important impact that automation may have on librarianship is that it presents an opportunity to invent the library of the future as though nothing existed today. This is reminiscent of J. C. R. Licklider's (1965) book *Libraries of the Future* in which he proposed the development of "procognitive systems" that would extend farther into the processes of generating, organizing, and using knowledge. Licklider cautioned that in thinking about procognitive systems, one must be prepared to reject the schema of the physical library.

Swanson (1967) concluded that "it is this notion of inventing, or planning, future libraries in the light of a critical re-examination of their purposes that may be regarded as having far-reaching implications for library education" (p. 2). This historical review of the impact of automation on public services suggests that this process of inventing in light of a critical re-examination has indeed begun. The challenge to all, both within and outside of library education, is to carry this process forward.

REFERENCES

Adkinson, B. W. (1964). Trends in library applications of data processing. In H. Goldhor (Ed.), *Proceedings of the 1963 clinic on library applications of data processing* (Papers presented at the 1st Annual Clinic, 28 April-1 May 1963) (pp.1-8). Urbana-Champaign: University of Illinois, Graduate School of Library Science.

American heritage dictionary of the English language. (1970). New York: Dell.

Anderson, C. R. (1989). Online ready reference in the public library. In L. C. Smith (Ed.), *Questions and answers: Strategies for using the electronic reference collection* (Papers presented at the 24th Annual Clinic on Library Applications of Data Processing, 5-7 April 1987) (pp. 71-84). Urbana-Champaign: University of Illinois, Graduate School of Library and Information Science.

Atkinson, H. C. (1972). The Ohio State On-line Circulation System. In F. W. Lancaster (Ed.), *Applications of on-line computers to library problems* (Papers presented at the 9th Annual Clinic on Library Applications of Data Processing, 30 April-3 May 1972) (pp. 22-28). Urbana-Champaign: University of Illinois, Graduate School of Library Science.

Bachelder, S. (1976). The New York Times Information Bank: A user's perspective. In F. W. Lancaster (Ed.), *The use of computers in literature searching and related reference activities in libraries* (Papers presented at the 12th Annual Clinic on Library Applications of Data Processing, 27-30 April 1975) (pp. 17-30). Urbana-Champaign: University of Illinois, Graduate School of Library Science.

Baker, B. (1989). Reference services and the networks: Some reflections on integration. In L. C. Smith (Ed.), *Questions and answers: Strategies for using the electronic reference collection* (Papers presented at the 24th Annual Clinic on Library Applications of Data Processing, 5-7 April 1987) (pp. 38-54). Urbana-Champaign: University of Illinois, Graduate School of Library and Information Science.

Beaumont, D. N. (1974). The LIBS 100 System. In F. W. Lancaster (Ed.), *Applications of minicomputers to library and related problems* (Papers presented at the 11th Annual Clinic on Library Applications of Data Processing, 28 April-1 May 1974) (pp. 55-79). Urbana-Champaign: University of Illinois, Graduate School of Library Science.

Becker, J. (1966). Current trends in library automation. In H. Goldhor (Ed.), *Proceedings of the 1966 clinic on library applications of data processing* (Papers presented at the 4th Annual Clinic, 24-27 April 1966) (pp. 1-6). Urbana-Champaign: University of Illinois, Graduate School of Library Science.

Becker, J. (1973). Library networks: The beacon lights. In F. W. Lancaster (Ed.), *Networking and other forms of cooperation* (Papers presented at the 10th Annual Clinic on Library Applications of Data Processing, 29 April-2 May 1973) (pp. 171-79). Urbana-Champaign: University of Illinois, Graduate School of Library Science.

Beckman, M. (1981). Public access at the University of Guelph library. In J. L. Divilbiss (Ed.), *Public access to library automation* (Papers presented at the 17th Annual Clinic on Library Applications of Data Processing, 20-23 April 1980) (pp. 37-48). Urbana-Champaign: University of Illinois, Graduate School of Library and Information Science.

Bingham, K. H., & Clark, B. M. (1989). The new CD-ROM technology: Shaping the future of reference and information research. In L. C. Smith (Ed.), *Questions and answers: Strategies for using the electronic reference collection* (Papers presented at the 24th Annual Clinic on Library Applications of Data Processing, 5-7 April 1987) (pp. 177-87). Urbana-Champaign: University of Illinois, Graduate School of Library and Information Science.

Briggs, R. B. (1976). The user interface for bibliographic search services. In F. W. Lancaster (Ed.), *The use of computers in literature searching and related reference activities in libraries* (Papers presented at the 12th Annual Clinic on Library Applications of Data Processing, 27-30 April 1975) (pp. 56-77). Urbana-Champaign: University of Illinois, Graduate School of Library Science.

Chapin, R. E. (1967). Administrative and economic considerations for library automation. In D. E. Carroll (Ed.), *Proceedings of the 1967 clinic on library applications of data processing* (Papers presented at the 5th Annual Clinic, 30 April-3 May 1967) (pp. 55-69). Urbana-Champaign: University of Illinois, Graduate School of Library Science.

Counts, R. W. (1967). Information services and operations of the Aerospace Research Applications Center (ARAC). In D. E. Carroll (Ed.), *Proceedings of the 1967 clinic on library applications of data processing* (Papers presented at the 5th Annual Clinic, 30 April-3 May 1967) (pp. 41-54). Urbana-Champaign: University of Illinois, Graduate School of Library Science.

Courtright, B. (1966). The Johns Hopkins University library. In H. Goldhor (Ed.), *Proceedings of the 1966 clinic on library applications of data processing* (Papers presented at the 4th Annual Clinic, 24-27 April 1966) (pp. 18-33). Urbana-Champaign: University of Illinois, Graduate School of Library Science.

Creth, S. (1989). Beyond technical issues: The impact of automation on library organizations. In L. C. Smith (Ed.), *Questions and answers: Strategies for using the electronic reference collection* (Papers presented at the 24th Annual Clinic on Library Applications of Data Processing, 5-7 April 1987) (pp. 4-13). Urbana-Champaign: University of Illinois, Graduate School of Library and Information Science.

Darling, L. (1966). Information retrieval projects in the Biomedical Library, University of California, Los Angeles. In H. Goldhor (Ed.), *Proceedings of the 1966 clinic on library applications of data processing* (Papers presented at the 4th Annual Clinic, 24-27 April 1966) (pp. 91-123). Urbana-Champaign: University of Illinois, Graduate School of Library Science.

Davison, W. (1974). Minicomputers and library automation: The Stanford experience. In F. W. Lancaster (Ed.), *Applications of minicomputers to library and related problems* (Papers presented at the 11th Annual Clinic on Library Applications of Data Processing, 28 April-1 May 1974) (pp. 80-95). Urbana-Champaign: University of Illinois, Graduate School of Library Science.

Diodato, V. (1989). Online ready reference in academic libraries: A survey of current practices and a review of planning issues. In L. C. Smith (Ed.), *Questions and answers: Strategies for using the electronic reference collection* (Papers presented at the 24th Annual Clinic on Library Applications of Data Processing, 5-7 April 1987) (pp.

55-70). Urbana-Champaign: University of Illinois, Graduate School of Library and Information Science.

Doszkocs, T. E. (1987). Natural language user interfaces in information retrieval. In F. W. Lancaster (Ed.), *What is user friendly?* (Papers presented at the 23rd Annual Clinic on Library Applications of Data Processing, 20-22 April 1986) (pp. 80-95). Urbana-Champaign: University of Illinois, Graduate School of Library and Information Science.

Dow, E. H. (1987). Personal information systems: The library role. *Library Journal, 112* (1 November), 29-32.

Dowlin, K. (1987). Aristotle meets Plato in the library catalog: Part 2. In F. W. Lancaster (Ed.), *What is user friendly?* (Papers presented at the 23rd Annual Clinic on Library Applications of Data Processing, 20-22 April 1986) (pp. 15-28). Urbana-Champaign: University of Illinois, Graduate School of Library and Information Science.

Evans, G. T. (1973). Bibliographic data centers for New York State. In F. W. Lancaster (Ed.), *Networking and other forms of cooperation* (Papers presented at the 10th Annual Clinic on Library Applications of Data Processing, 29 April-2 May 1973) (pp. 150-64). Urbana-Champaign: University of Illinois, Graduate School of Library Science.

Goldhor, H. (1964). Foreword. In H. Goldhor (Ed.), *Proceedings of the 1963 clinic on library applications of data processing* (Papers presented at the 1st Annual Clinic, 28 April-1 May 1963) (pp. iii-v). Urbana-Champaign: University of Illinois, Graduate School of Library Science.

Griffin, M. (1964). IBM advanced systems development library in transition. In H. Goldhor (Ed.), *Proceedings of the 1963 clinic on library applications of data processing* (Papers presented at the 1st Annual Clinic, 28 April-1 May 1963) (pp. 79-95). Urbana-Champaign: University of Illinois, Graduate School of Library Science.

Harter, S. P. (1989). Online searching as a problem solving process. In L. C. Smith (Ed.), *Questions and answers: Strategies for using the electronic reference collection* (Papers presented at the 24th Annual Clinic on Library Applications of Data Processing, 5-7 April 1987) (pp. 103-20). Urbana-Champaign: University of Illinois, Graduate School of Library and Information Science.

Hewett, T. T., & Meadow, C. T. (1981). A study of the measurement of user performance. In J. L. Divilbiss (Ed.), *Public access to library automation* (Papers presented at the 17th Annual Clinic on Library Applications of Data Processing, 20-23 April 1980) (pp. 49-76). Urbana-Champaign: University of Illinois, Graduate School of Library and Information Science.

Hildreth, C. R. (1989). Extending the access and reference service capabilities of the online public access catalog. In L. C. Smith (Ed.), *Questions and answers: Strategies for using the electronic reference collection* (Papers presented at the 24th Annual Clinic on Library Applications of Data Processing, 5-7 April 1987) (pp. 14-33). Urbana-Champaign: University of Illinois, Graduate School of Library and Information Science.

Kennedy, R. A. (1970). Bell Laboratories on-line circulation control system: One year's experience. In D. E. Carroll (Ed.), *Proceedings of the 1969 clinic on library applications of data processing* (Papers presented at the 7th Annual Clinic, 27-30 April 1969) (pp. 14-30). Urbana-Champaign: University of Illinois, Graduate School of Library Science.

Kessler, M. M. (1966). The Technical Information Project of the Massachusetts Institute of Technology. In H. Goldhor (Ed.), *Proceedings of the 1966 clinic on library applications of data processing* (Papers presented at the 4th Annual Clinic, 24-27 April 1966) (pp. 7-17). Urbana-Champaign: University of Illinois, Graduate School of Library Science.

Kilgour, F. G. (1977). The economics of library computerization. In J. L. Divilbiss (Ed.), *The economics of library automation* (Papers presented at the 13th Annual Clinic on Library Applications of Data Processing, 25-28 April 1976) (pp. 3-9). Urbana-Champaign: University of Illinois, Graduate School of Library Science.

Kilgour, F. G. (1969). Initial system design for the Ohio College Library Center: A case history. In D. E. Carroll (Ed.), *Proceedings of the 1968 clinic on library applications of data processing* (Papers presented at the 6th Annual Clinic, 5-8 May 1968) (pp. 79-88). Urbana-Champaign: University of Illinois, Graduate School of Library Science.

Kraft, D. H. (1964). Application of IBM equipment to library mechanization, keyword-in-context (KWIC) indexing and the selective dissemination of information (SDI). In H. Goldhor (Ed.), *Proceedings of the 1963 clinic on library applications of data processing* (Papers presented at the 1st Annual Clinic, 28 April-1 May 1963) (pp. 133-56). Urbana-Champaign: University of Illinois, Graduate School of Library Science.

Kubitz, W. J. (1980). Computer technology: A forecast for the future. In F. W. Lancaster (Ed.), *The role of the library in an electronic society* (Papers presented at the 16th Annual Clinic on Library Applications of Data Processing, 22-25 April 1979) (pp. 135-61). Urbana-Champaign: University of Illinois, Graduate School of Library Science.

Lancaster, F. W. (1976). Have information services been successful? A critique. In F. W. Lancaster (Ed.), *The use of computers in literature searching and related reference activities in libraries* (Papers presented at the 12th Annual Clinic on Library Applications of Data Processing, 27-30 April 1975) (pp. 145-56). Urbana-Champaign: University of Illinois, Graduate School of Library Science.

Licklider, J. C. R. (1965). *Libraries of the future.* Cambridge, MA: MIT Press.

Lipetz, B.-A. (1970). A quantitative study of catalog use. In D. E. Carroll (Ed.), *Proceedings of the 1969 clinic on library applications of data processing* (Papers presented at the 7th Annual Clinic, 27-30 April 1969) (pp. 42-49). Urbana-Champaign: University of Illinois, Graduate School of Library Science.

McCarn, D. B. (1973). Network—Or all hang separately. In F. W. Lancaster (Ed.), *Networking and other forms of cooperation* (Papers presented at the 10th Annual Clinic on Library Applications of Data Processing, 29 April-2 May 1973) (pp. 1-8). Urbana-Champaign: University of Illinois, Graduate School of Library Science.

Meadow, C. T. (1989). Tailoring system design to users. In L. C. Smith (Ed.), *Questions and answers: Strategies for using the electronic reference collection* (Papers presented at the 24th Annual Clinic on Library Applications of Data Processing, 5-7 April 1987) (pp. 121-31). Urbana-Champaign: University of Illinois, Graduate School of Library and Information Science.

Mischo, W. H., & DeSart, M. G. (1989). An end user search service with customized interface software. In L. C. Smith (Ed.), *Questions and answers: Strategies for using the electronic reference collection* (Papers presented at the 24th Annual Clinic on Library Applications of Data Processing, 5-7 April 1987) (pp. 188-204). Urbana-Champaign: University of Illinois, Graduate School of Library and Information Science.

Negus, A. E. (1974). The application of minicomputers to problems of information retrieval. In F. W. Lancaster (Ed.), *Applications of minicomputers to library and related problems* (Papers presented at the 11th Annual Clinic on Library Applications of Data Processing, 28 April-1 May 1974) (pp. 96-104). Urbana-Champaign: University of Illinois, Graduate School of Library Science.

Nitecki, D. A. (1983). Competencies required of public services librarians to use new technologies. In L. C. Smith (Ed.), *Professional competencies—technology and the librarian* (Papers presented at the 20th Annual Clinic on Library Applications of Data Processing, 24-26 April 1983) (pp. 43-57). Urbana-Champaign: University of Illinois, Graduate School of Library and Information Science.

Noreault, T. (1989). Optical publishing: Effects on reference services. In L. C. Smith (Ed.), *Questions and answers: Strategies for using the electronic reference collection* (Papers presented at the 24th Annual Clinic on Library Applications of Data Processing, 5-7 April 1987) (pp. 94-102). Urbana-Champaign: University of Illinois, Graduate School of Library and Information Science.

Oliver, D. (1985). Electronic mail services in the library and information center community. In J. L. Divilbiss (Ed.), *Telecommunications: Making sense of new technology and new legislation* (Papers presented at the 21st Annual Clinic on Library Applications of Data Processing, 14-16 April 1984) (pp. 77-83). Urbana-Champaign: University of Illinois, Graduate School of Library and Information Science.

Parker, R. H. (1964). Development of automatic systems at the University of Missouri library. In H. Goldhor (Ed.), *Proceedings of the 1963 clinic on library applications of data processing* (Papers presented at the 1st Annual Clinic, 28 April-1 May 1963) (pp. 43-55). Urbana-Champaign: University of Illinois, Graduate School of Library Science.

Paulukonis, J. T. (1972). On-line real-time self-service circulation at Northwestern University. In F. W. Lancaster (Ed.), *Applications of on-line computers to library problems* (Papers presented at the 9th Annual Clinic on Library Applications of Data Processing, 30 April-3 May 1972) (pp. 82-93). Urbana-Champaign: University of Illinois, Graduate School of Library Science.

Pizer, I. H. (1972). On-line technology in a library network. In F. W. Lancaster (Ed.), *Applications of on-line computers to library problems* (Papers presented at the 9th Annual Clinic on Library Applications of Data Processing, 30 April-3 May 1972) (pp. 54-68). Urbana-Champaign: University of Illinois, Graduate School of Library Science.

Price, D. de S. (1980). Happiness is a warm librarian. In F. W. Lancaster (Ed.), *The role of the library in an electronic society* (Papers presented at the 16th Annual Clinic on Library Applications of Data Processing, 22-25 April 1979) (pp. 3-15). Urbana-Champaign: University of Illinois, Graduate School of Library Science.

Randall, G. E. (1973). Interlibrary cooperation in an industrial environment. In F. W. Lancaster (Ed.), *Networking and other forms of cooperation* (Papers presented at the 10th Annual Clinic on Library Applications of Data Processing, 29 April-2 May 1973) (pp. 113-23). Urbana-Champaign: University of Illinois, Graduate School of Library Science.

The Random House dictionary, concise edition. (1980). New York: Random House.

Shera, J. (1964). Automation and the reference librarian. *RQ, 3*(6), 3-7.

Smith, L. C. (Ed.). (1989). *Questions and answers: Strategies for using the electronic reference collection* (Papers presented at the 24th Annual Clinic on Library Applications of Data Processing, 5-7 April 1987). Urbana-Champaign: University of Illinois, Graduate School of Library and Information Science.

Summit, R. K., & Drew, S. J. (1976). The pubic library as information dissemination center: An experiment in information retrieval services for the general public. In F. W. Lancaster (Ed.), *The use of computers in literature searching and related reference activities in libraries* (Papers presented at the 12th Annual Clinic on Library Applications of Data Processing, 27-30 April 1975) (pp. 91-102). Urbana-Champaign: University of Illinois, Graduate School of Library Science.

Swanson, D. R. (1967). Education and library automation. In D. E. Carroll (Ed.), *Proceedings of the 1967 clinic on library applications of data processing* (Papers presented at the 5th Annual Clinic, 30 April-3 May 1967) (pp. 1-7). Urbana-Champaign: University of Illinois, Graduate School of Library Science.

Sweeney, R. T. (1983). The public librarian of the last years of the twentieth century. In L. C. Smith (Ed.), *Professional competencies—Technology and the librarian* (Papers presented at the 20th Annual Clinic on Library Applications of Data Processing, 24-26 April 1983) (pp. 58-68). Urbana-Champaign: University of Illinois, Graduate School of Library and Information Science.

Taine, S. I. (1964). Bibliographic data processing at the National Library of Medicine. In H. Goldhor (Ed.), *Proceedings of the 1963 clinic on library applications of data processing* (Papers presented at the 1st Annual Clinic, 28 April-1 May 1963) (pp. 109-32). Urbana-Champaign: University of Illinois, Graduate School of Library Science.

Toliver, D. E. (1987). Design issues in automatic translation for online information retrieval systems. In F. W. Lancaster (Ed.), *What is user friendly?* (Papers presented at the 23rd Annual Clinic on Library Applications of Data Processing, 20-22 April 1986) (pp. 96-107). Urbana-Champaign: University of Illinois, Graduate School of Library and Information Science.

Turoff, M., & Hiltz, S. R. (1980). Electronic information exchange and its impact on libraries. In F. W. Lancaster (Ed.), *The role of the library in an electronic society* (Papers presented at the 16th Annual Clinic on Library Applications of Data Processing, 22-25 April 1979) (pp. 117-34). Urbana-Champaign: University of Illinois, Graduate School of Library Science.

Waltz, D. L. (1976). Natural-language question-answering systems. In F. W. Lancaster (Ed.), *The use of computers in literature searching and related reference activities in libraries* (Papers presented at the 12th Annual Clinic on Library Applications of Data Processing, 27-30 April 1975) (pp. 137-44). Urbana-Champaign: University of Illinois, Graduate School of Library Science.

Wanat, C. (1985). Management strategies for personal files: The Berkeley seminar. *Special Libraries, 76*(Fall), 253-60.

Warheit, I. A. (1972). On-line interactive systems in libraries, now and in the future. In F. W. Lancaster (Ed.), *Applications of on-line computers to library problems* (Papers presented at the 9th Annual Clinic on Library Applications of Data Processing, 30 April-3 May 1972) (pp. 3-21). Urbana-Champaign: University of Illinois, Graduate School of Library Science.

Waterman, D. A. (1986). *A guide to expert systems.* Reading, MA: Addison-Wesley.

Wax, D. M. (1976). NASIC and the information services librarian: Room in the middle. In F. W. Lancaster (Ed.), *The use of computers in literature searching and related reference activities in libraries* (Papers presented at the 12th Annual Clinic on Library Applications of Data Processing, 27-30 April 1975) (pp. 78-90). Urbana-Champaign: University of Illinois, Graduate School of Library Science.

Webster's new collegiate dictionary. (1956). Springfield, MA: Merriam.

Wertz, J. A. (1965). Possible applications of data processing equipment in libraries. In H. Goldhor (Ed.), *Proceedings of the 1964 clinic on library applications of data processing* (Papers presented at the 2nd Annual Clinic, 26-29 April 1964) (pp. 112-17). Urbana-Champaign: University of Illinois, Graduate School of Library Science.

Wigington, R. L. (1982). Technology alone is not enough. In L. C. Smith (Ed.), *New information technologies—New opportunities* (Papers presented at the 18th Annual Clinic on Library Applications of Data Processing, 26-29 April 1981) (pp. 3-12). Urbana-Champaign: University of Illinois, Graduate School of Library and Information Science.

Williams, M. E. (1980). Future directions for machine-readable data bases and their use. In F. W. Lancaster (Ed.), *The role of the library in an electronic society* (Papers presented at the 16th Annual Clinic on Library Applications of Data Processing, 22-25 April 1979) (pp. 82-93). Urbana-Champaign: University of Illinois, Graduate School of Library Science.

Woodard, B. S. (1989). Strategies for providing public service with an online catalog. In L. C. Smith (Ed.), *Questions and answers: Strategies for using the electronic reference collection* (Papers presented at the 24th Annual Clinic on Library Applications of Data Processing, 5-7 April 1987) (pp. 71-84). Urbana-Champaign: University of Illinois, Graduate School of Library and Information Science.

KATHRYN LUTHER HENDERSON*

Professor
Graduate School of Library and Information Science
University of Illinois at Urbana-Champaign

and

WILLIAM T HENDERSON

Preservation Librarian
Associate Professor of Library Administration
University of Illinois at Urbana-Champaign

From Flow Charting to User Friendly: Technical Services Functions in Retrospect

INTRODUCTION

In this presentation, the proceedings of the twenty-four preceding Clinics on Library Applications of Data Processing will be summarized to give a flavor of the issues and themes relating to technical services functions which have been defined, for purposes of this paper, as: acquisitions, serials control and management, and catalogs and cataloging.

In 1963, the world was waking up to a new era of technology influencing many aspects of life. Increased technology was reported to be costing workers their jobs and causing labor unrest. Cited as evidence was the fact that on February 11, 1963, eleven electronic computers took over the jobs of the many people required to tabulate stock market figures in New York for the nationwide wires of the Associated Press (*Year,* 1963, p. 25). Gordon Cooper was the last of the Project Mercury astronauts to go into orbit. After a successful day and a half in space, the spacecraft's automatic controls went dead, but Cooper landed safely (p. 28). At the University of Illinois on March 2, 1963, the spaceship-

* The authors contributed equally to this publication. Names are listed in alphabetical order.

shaped Assembly Hall was dedicated—not only did its shape reflect the times, but it was one of the first buildings in the country to make use of sophisticated computer controls (Thomas Parkinson to Rebecca Hall, WCIA, Channel 3 broadcast, Champaign, Illinois, 5 March 1988). *The New International Yearbook for the Year 1963* heralded the development of thirty new commercial digital computer models, most impressive of which was the Control Data Corporation's 6600 with a central memory of 131,000 60-bit words, exceeding in speed and memory capacity all available computers. Noteworthy, too, was a new computer language, FORTRAN IV (*The New International Yearbook for the Year 1963,* 1964, p. 115).

Significant events also occurred in the library world. At the September 1962 faculty meeting of the Graduate School of Library Science, Herbert Goldhor, Director, had sought and received approval for the School to hold a Clinic on Library Applications of Data Processing in the spring of 1963 (University of Illinois, 1962). Goldhor had come to the School a few months earlier after serving for a decade as director of the Evansville (Indiana) Public Library. He was impressed with the possibilities of applications of mechanization to library operations and felt that the School had a responsibility to foster programs of continuing education along that line (Goldhor, personal communication, February 29, 1988). By March 28, 1963, the faculty was informed that 123 applications were received for the first Clinic, and 92 were selected to attend (University of Illinois, 1963).

A conference entitled "Libraries and Automation" sponsored by the Library of Congress, the National Science Foundation, and the Council on Library Resources was held at Airlie Foundation, Warrenton, Virginia from May 26-30, 1963. One hundred people who were planning or who had mechanization projects underway in research libraries were invited to attend (Markuson, 1964). In 1963, *Automation and the Library of Congress* was published; it became better known as the King Report, after Gilbert W. King (1963), chairperson of the study. It, too, was sponsored by the Council on Library Resources.

In reviewing the work of technical services in 1963, Maurice F. Tauber (1964), writing in *Library Resources and Technical Services,* remarked "that documentation has been making a marked impression upon more and more librarians" (p. 104). Two articles in *LRTS* in 1963 were devoted to library automation (Richmond, 1963; Fasana, 1963)

The 1963 volume of *College and Research Libraries* contained only one major article that dealt with data processing. This article summarized the July 18, 1963 membership meeting of ACRL's University Libraries Section. In the article, Don S. Culbertson, Melvin J. Voigt and James R. Cox discussed data processing costs in acquisitions, cataloging, serials, and circulation (Culbertson et al., 1963, pp. 487-95). In his summary of acquisitions and cataloging, Culbertson concluded by noting that James Skipper, when he accepted the gavel as president of the Resources and Technical Services Division, had indicated that "the future belongs to technical services." Culbertson thought Skipper to be right. "Unless", remarked Culbertson, "we get our costs under control and become able to keep them there, the whole library risks belonging to technical services" (p. 489). Little wonder, then, that in reviewing the first Illinois Clinic, Carl R. Cox (1965) acknowledged that, for the most part, "all applications of data processing to date have been in the area of the library's housekeeping operations, technical services, and circulation" (p. 409). The Clinic's early presentations concentrated on technical services operations. The conferences were first intended to present papers from librarians from various types of libraries with sufficient relevant experience in the mechanization of library operations to make their findings of value to others through case reports of their experiences (Goldhor, personal communication, February 29, 1988). And Goldhor noted in the "Foreword" to the first volume, the strong emphasis was on "routines" with little emphasis on information retrieval and none on information storage (Goldhor, 1964, p. iv).

SYSTEMS APPROACH

The routines of library processes, as Burton W. Adkinson (1963) remarked in the keynote paper of the first conference, were not simple problems to solve in machine systems. He noted the trend toward realizing that "a library performs many services which can best be approached from a systems point of view" (p. 2). This approach assumed that machine records in one operation, for example, acquisitions, could be used for cataloging, circulation, selective dissemination, or information retrieval. Adkinson saw this approach as sending librarians back to "first principles" and studying in detail the basic input record for each item in the library (p. 2) as each library built its own stand-alone system.

The systems approach required an understanding of an existing system, not to mimic that system, but to be sure that all parts of the process were taken into account. A relatively new tool to librarians, but one which could become the link between them and the "machine people," was the flow chart designed to be "a graphic representation of a procedure's flow, showing the decisions that need to be made and the actions that must be taken to complete a particular task or series of tasks" (Schultheiss, 1962, p. 79). The sequence in which tasks were to be carried out was another important feature of flow charting.

Louis Schultheiss (1964) introduced the first Clinic's audience to flowcharting, a device that would become an integral part of the papers of many Clinics to follow, particularly in regard to technical service activities. Not only was it seen as a method to advance the systems approach and define the process for "machine people," but, as Jane Burke (1986) indicated in her Clinic presentation twenty-two years later, flowcharting was a good way to involve all levels of the staff in the human aspects of library automation because it is a "group process" that holds people together and facilitates communication at different levels (p. 51). Thus flowcharting has endured to facilitate both human and technical activities.

Acquisitions

Many of the early efforts in automating acquisitions were part of the early system approaches. At the first Clinic in 1963, five papers included material on the automation of the ordering of books, with three of them indicating that information generated in the order process was used also for cataloging processes. In these early years of the Clinic, it was not uncommon for a system to have begun as either an ordering system or a cataloging system and subsequently to have had the other activity added to it. Along with generating orders and creating lists of materials on order and lists of materials in process, a number of these early systems also handled billing and payment operations, printed labels and book cards, and performed other routine tasks. Approximately one-third of the presentations made in the first three Clinics, 1963-65, included coverage of automation activity relating to acquisitions. Though small in numbers compared to the entire population of librarians, there were librarians at that time trying out and using data processing in very sophisticated ways such as Lorin R. Burns (1964) reporting on the acquisitions system of the Lake County Indiana Public Library; Ralph Parker (1964), reporting on the University of Missouri Library; Hillis Griffin (1964), on the National Reactor Testing Station Technical Library; and Walter Curley (1966), on the Suffolk Cooperative Library. An interesting sidelight on this is that a book jobber, in relating his firm's experiences in automating, indicated that, during this period, there was need for special programming in order to provide for the interchange of data with libraries using data processing (Brody, 1966).

During the later 1960s, several libraries reported on other interesting and significant developments in the automation of acquisitions as part of planned integrated systems. One of them was included in Charles T. Payne's (1967) description of the genesis of the bibliographic data system of the University of Chicago Library. In this system, all activity relating to a title was linked to a single record which served for controlling all aspects of the use and treatment of the title. At the time of Payne's report, the system was not fully operational and his presentation dealt with identification of bibliographical data elements and their manipulation in the system, and acquisitions was not his major emphasis. A subsequent report on the Chicago system made in 1970, at the beginning of the MARC era, included a section on acquisitions (McGee & Miller, 1971, pp. 78-97).

Another interesting integrated system known by the acronym LISTS, was introduced in Los Angeles in 1968 by the System Development Corporation (Black, 1969). Used by the technical library of the System Development Corporation and a limited number of libraries in the Los Angeles area, this system was developed to determine costs of operating online systems and to determine acceptable cost levels for such systems. It was based on MARC records, and though not fully operational at the time of the report, was intended eventually to perform ordering and order control functions together with cataloging and circulation, as well as being designed to produce a great deal of management data on all phases of its operation.

Washington State University also was developing an integrated system, and, as described at the 1969 Clinic, its acquisitions subsystem was not yet operating, but a number of criteria for it had been identified and were outlined in the report as including: collecting necessary order data at the beginning; provision for effective file management and immediate and continuing updating of information in the files; utilizing both online and batch processing; immediate inquiry from many access points; provision of as much automatic processing as possible (e.g., automated editing of old outstanding orders and automatic removal from the file when materials were no longer in processing); development of management information on the acquisition process; flexibility of procedures; simplicity of operation; and potential for expansion to handle increasing numbers of orders (Burgess, 1970).

The linking of acquisitions with cataloging continued as a pattern throughout the latter half of the 1960s, but these years also brought a limited number of full-length presentations describing acquisitions systems in detail. The first of these long reports is in the 1968 Proceedings and describes an automated order system in operation at the University of Michigan (Thomson & Muller, 1969). The report, made after the system had been in operation for two years, described the care with which the manual system and the library's needs for information on its acquisitions activity had been analyzed and flowcharted before any programming had been done. The system, based on punched cards and batch processing, had proven itself in use, had absorbed increased order activity, had controlled staff growth in the acquisitions unit so that additional space was not required, and was judged successful on other counts as well. One interesting conclusion made at Michigan was that comparing costs of operating this system for one year with another was difficult because the amount of activity varied too greatly from one year to another for such comparisons to be valid.

The Georgia Tech Library, in the late 1960s, reported that it planned to expand its MARC-based cataloging system into other areas, including book selection by generating, from each week's new MARC tape, printouts of records in a particular classification area for each of a corps of selectors (Kennedy, 1969). Ultimately this library intended that its system would include selection, ordering, and cataloging all based on the MARC records. The idea of breaking out segments of the records in particular LC classification segments harks back to the preautomation use of LC proof slips for selection and ordering purposes.

Thus, with the appearance of MARC and the proliferation of catalog projects which were spun off it, acquisitions almost disappears from the proceedings of these Clinics except for a pair of papers in 1972 describing online systems called LOLITA and BALLOTS. Upon reading the account of LOLITA's origin and operation, it appears to have been a friendly and helpful adjunct to the Oregon State University Library's acquisitions program (Auld & Baker, 1972). LOLITA was, of course, an acronym for a much longer and less catchy name for the system which was an online book order and fund accounting system that had been in operation for two years. It automatically indexed in an overnight operation those orders input during the previous day so that they thereafter were searchable by author as well as by order number. As invoice information was input each day, it was automatically linked to the original order record, and there was provision in a special file for entry and control of invoice data relating to subscriptions, standing orders, and binding, a feature which enabled the acquisitions unit to consolidate all its accounting work in the system. LOLITA was indigenous to the Oregon State campus and its terminals could be operated only within a very limited distance of the OSU computer center; however, because others were becoming interested in it, there was interest in freeing LOLITA of these constraints. LOLITA was considered to be a success. It existed in a dynamic environment and continued to benefit from continual change and improvement.

BALLOTS, which originated at Stanford University, differed from LOLITA in that it was designed from the beginning to be a total integrated system and it was so described in a report by A. H. Epstein and Allen B. Veaner (1972). At that time, the system was still being developed, and the report described the acquisitions module which was the first of eleven projected segments to be programmed. In the system, MARC records stored on tape and updated weekly were extensively indexed to permit ready online searching by terminal operators. Inputting author or author/title information enabled one to initiate an order after which the system edited the information which had been entered, indicated errors and omissions, and, finally, accepted the complete and corrected record. At that time the operator could proceed with the entry of another order. Orders were printed overnight for dispatch to vendors. The then–novel idea of the cursor to mark the operator's progress was described in some detail as was the use of special keys on the terminal for special purposes. This latter feature was an early manifestation of function keys now so familiar to all using personal computers.

Serials Control and Management

Serial control is a recurring topic in many papers presented, particularly in the initial decade of the Clinics. It crops up in connection with plans for integrated systems and in individual libraries' lists of areas which had been or which were to be automated; however, few institutions or individuals made full-scale reports of the development of serial control systems. Parker (1964), in the first Clinic in 1963, indicated that machine control of subscription records at the University of Missouri Library was an accomplished fact; and, in that same conference, two separate reports on automation efforts in different libraries indicated the use of data processing techniques to route periodicals to patrons and to prepare periodical lists (Griffin, 1964; and H. Griffin, 1964). Proceedings of the first conference also contain one of the long papers detailing automated work with serials in Seymour I. Taine's account, in 1963, of the development of the indexing programs of the National Library of Medicine (NLM) (Taine, 1964). While not a serial control system, the early development of automated systems at NLM for its index publication program was a pioneering effort in the use of automation in libraries and gave results which are now recognized both as basic to the growth of information services in medicine and of library applications of data processing in general. These developments at NLM also are indications of what can transpire if both planning for and funding of library automation are adequate and consistent.

In subsequent years serial holdings lists and serial control systems continued to be included in descriptions of automation projects in particular libraries or were mentioned as projects under development or still in the planning stages by those reporting on the development of comprehensive systems. A short description of a serial control system within a longer paper outlines the use of a minicomputer as the base for the development of an online integrated system in the University of Minnesota's Bio-Medical Library (Brudvig, 1974). This is actually a manifestation of a later era incorporating much more sophisticated equipment than was in use in the early Clinic years.

Robert Kozlow (1968) reported on the serial holdings list of the University of Illinois at Urbana-Champaign Library. This is the only paper in the proceedings devoted wholly to the generation of serial holdings lists and contains a detailed description of the techniques, based on punched cards, utilized in publishing a truly large listing of this type.

Conferees in 1972 heard a detailed report on the development and operation of the "Terminal Oriented Real Time Operating System (TORTOS), an online serial control system at the UCLA Bio-Medical Library (Fayollat, 1972). This system was designed to eliminate card handling and to keep the database current on a daily basis. It was reported to be cost–effective and to be providing measurable advantages over a previous batch system. The system was still under development at the time of the report and there were plans to include processing of invoice information in addition to controlling holdings.

The last full–length presentation relating specifically to serials was in 1973, when Charles Sage reported on the utilization of the MARC II format for serials in standardizing the handling of machine-readable serials records by the three state university libraries in Iowa (Sage, 1974). This effort succeeded in converting all three sets of serial data to a common format and in producing lists of the serial holdings of each of the three libraries using the MARC formatted records, but it did not succeed in producing a union list without the intervention of human workers to match titles appearing in more than one of the individual databases.

This partial success ends this account of the history of serial data automation as reported in these Clinics. By the early 1970s the utilization of MARC records, the beginnings of work on making online catalogs, and the coverage of other themes in the Clinics had assumed such proportions that the automation of serial records and acquisitions procedures disappeared from the agendas of the conference. In general, acquisitions and serials systems in the online environment have not become as prolific as might have been expected. In a recent study of 209 four-year academic institutions in the United States, it was found that 169 (80.9 percent) did not have online acquisitions systems; of the 41 that did, 20 used a system supplied through a bibliographic utility. Only twenty-six (12.5 percent) had an online serials control system with no single type predominating (Camp et al., 1987, pp. 341-42).

(These findings are corroborated by a 1986 survey conducted by the Association of American Publishers/American Library Association Resources and Technical Services Division Joint Committee which found that few of the wholesalers provide electronic ordering systems for use by libraries (Edelman & Muller, 1987). In addition, Marcia Tuttle has noted that serials acquisition had been considered a likely place to begin automation activities but that "systems designers did not understand serials; the high incidence of exception to the rule made automation of serials receipt too difficult for existing technology. Thus, most data processors turned to cataloging and other library functions" [Derthicke & Moran, 1986, p. 14]).

In reviewing the past quarter century, Jim Segesta and Rod Hersberger (1987) note the revolution in automated cataloging and reference services through OCLC, BRS, Dialog, SDC, etc. and remark that collaboration between the book trade and libraries has not materialized. The integrated systems approach diminished over the years. Gerard Salton (1980) would speak to the impressiveness of some integrated systems while acknowledging that "the recent trend in the direction of cooperative ventures among libraries has somewhat damaged the enthusiasm for the integrated stand-alone systems, and the feeling now seems to be that they are too costly to be supported by single library organizations without substantial outside aid" (p. 61). Michael Gorman (1987) concluded that there were few, if any, truly integrated systems in medium sized or large libraries although many libraries had created a hybrid of partially integrated and partially separate systems. The choice has always seemed to be "between the complex architecture of the integrated system and the user hostility of the separate system approach" with the microcomputer making possible a third choice (p. 4). Perhaps in another twenty-five years truly integrated systems will have come into their own, and acquisitions and serial control may have again become familiar themes in the literature of library automation.

Catalogs and Cataloging

Format

Mechanized and computerized means affected cataloging and catalogs in several important ways. Three ways will be emphasized here: the format of the catalog, the methods of distributing and the degree of sharing cataloging data, and the catalog users. All of these are intertwined in the collage that will be developed from the pieces presented in the Clinic's proceedings.

The pieces began to form in the 1950s and 1960s when, with the advent of better mechanized means promising to produce a much higher quality product (Stromberg, 1966, p. 196), libraries reinvented an old format, the book catalog, in an attempt to cut costs and provide better service. Remembered were many advantages of the book catalog (Stromberg, 1966); seemingly forgotten were disadvantages which did not disappear. Among the reasons cited for this movement back to the book catalog after seventy years of card catalog reign were cost factors such as elimination of the preparation of card catalogs for multiple locations and the maintenance of card files. The freeing of prime space occupied by card catalog cases was an extra benefit as was the saving of staff time coming from the elimination of catalog maintenance costs for typing cards and headings, sorting, filing, and reviewing. And who could quarrel with increased service that multiple copies could provide for all internal services to staff and users as well as external service to branches, other libraries, and users in their homes and offices? In addition, advancements in technology made for a much higher quality product.

Faced with providing service to a burgeoning population and the opening of new branches without catalogs, the Los Angeles County Public Library, in 1952, turned to the production of book catalogs. A report on the Los Angeles catalogs was made by John D. Henderson (1964) at the first Clinic. Like many of the early Clinic papers, this one went into great detail about basic designs of tabulating cards and the equipment used. Four years later the Clinic included another paper chronicling the progress of the Los Angeles book catalogs from unit record (punched card) equipment, to sequential camera processing, to preparations for conversion to a computerized system. This was indicative of the changes that were occurring in technology (Zuckerman, 1968).

Other early Clinics reported the changing means for mechanized book catalog production in all types of libraries. Even by 1976, Maurice J. Freedman (1977) was having a difficult time giving up the book catalog, although by then, he was admitting that the book format was more costly and more quickly outdated than its newer format rivals such as the Computer-Output-Microform (COM) catalog (p. 120) which was enjoying a brief interlude in the format limelight on the way to the online catalog and which was the subject of a report at the same Clinic (Maliconico, 1977).

The physical format of the catalog has been noted but the word *format* took on a new meaning in the machine era. Everyone knew that the bibliographic content of records could vary considerably, but as libraries developed stand-alone mechanized systems, there were other variations in machine-readable records. Ann Curran (1970) summarized some of these variations including the recording of the data (e.g., Is it in natural language, coded form, or both?); the determination of items that are to be separately identified in the record; the manner of identification of these items (as tagged fields, subfields, etc.); the structure of the machine records (Are the identifying tags interspersed with the data?); and the character set used to represent the data (p. 38). Lack of standardization in these areas precluded the ready exchange of bibliographic data in machine-readable form among libraries. There were some cooperative efforts to produce technical services products reported at the early Clinics, such as the Harvard-Columbia-Yale medical libraries project (Kilgour, 1965), the Suffolk Cooperative Library System (Curley, 1966), the Los Angeles school systems (Dodendorf, 1966), and the Ontario New Universities Library Project (Bregzis, 1966).

To facilitate the development of a national library system, the 1963 King Report had emphasized the conversion of information from textual form to machine-readable form as a necessary step for both the transition to an automated system and its continual updating; but, at that date, team members admitted that they did not know the techniques to be used in the conversion (King et al., 1963, p. 9); by 1967, when Barbara Evans Markuson (then Chief, Library Methods Staff, Information Systems Office, Library of Congress, who had edited the 1963 Airlie Conference proceedings and assisted the King Report team) introduced MARC into the Clinic proceedings, a few volunteer libraries had been receiving machine-readable data in the MARC I communications format from the Library of Congress since November 1966 (Markuson, 1968, p. 106). She saw the new system as one likely to influence the future characteristics of the national library system or network (pp. 111-12). And she was correct, for with MARC, a whole new era of library automation had begun and one would have to look hard to find a Clinic after 1967 that did not include that now-familiar acronym somewhere in its pages. At the next year's Clinic, John P. Kennedy (1969) described Georgia Institute of Technology's use of the MARC Pilot Project tapes to produce catalog cards and book catalogs as a highly successful experiment that increased production; but he also noted the expensive development costs, the difficulty of obtaining programmers, the lack of documentation, and the arduous process of testing and debugging (pp. 199-215).

The Library of Congress continued work on the format that faciliated the machine-readable interchange of bibliographic data between institutions and, in 1969, began to distribute, through the MARC Distribution Service, its cataloging services on machine-readable tape. Eager to report how libraries were using these services, the planners of the 1970 Clinic departed from the past consideration of all major aspects of library operations in a particular Clinic to concentrate on one theme, MARC, which had implications for many different library operations.

(Using a special theme for each Clinic has continued ever since.) The 1970 Clinic's intent "was to review MARC after one year of operation, to present the current picture and future programs of the Library of Congress in regard to MARC and to assess the local, national, and international potential of this service" (p. ix).

Henrietta D. Avram (1971) spoke on the purpose of a data-oriented, computer-based centralized service with emphasis toward generalized applications on a centrally maintained set of data files for access by a variety of users. She saw the evolving MARC system as part of this data utility. Hillis Griffin (1971), a distributor of the MARC tapes to local libraries and a MARC user, praised some users for their hard work but chided other subscribers for lack of initiative in using MARC. By 1970, MARC was "international" and the British MARC program was described by R. E. Coward (1971). Other library and commercial users of MARC rounded out the group. In general, the users characterized MARC as being timely, sophisticated, difficult to manipulate because it had never been done before, eminently usable, and requiring hard work. The concept of distributing data in this fashion was viewed as being so radical as to require rethinking of each individual library operation as part of a total system because each aspect of library administration would be affected by MARC (McGee, 1971, p. 96). Most speakers shared Fred Kilgour's (1971) optimism for the future of library automation which he evoked in the conference's closing presentation. As Michael Gorman (1977) discussed conversion of existing bibliographic files into machine-readable form, he presumed that the format in which the records would be received would be the MARC format—or at least a MARC-compatible format (p. 125). In a short space of time, MARC had, indeed, become the standard and changed much of the direction of library automation.

Methods of Distributing and the Degree of Sharing Cataloging Data: Networking

One of the problems in using machine-readable data distributed in MARC format by the Library of Congress was that most local libraries didn't know what to do with the data once they had it, and/or they did not have the expertise or the equipment to process it. Library rumor has it that some purchased the tapes merely as a status symbol—to be able to answer affirmatively when, at library conferences, one was asked "Are you getting MARC tapes?" and then quickly remember another appointment before being asked "What are you doing with them?" Too often the answer to that would have been "putting them on a shelf."

Would cooperative efforts be a way for libraries to process and use the data? At the 1968 Clinic, Frederick G. Kilgour (1969) reported on the Ohio College Library Center (OCLC), a project that he made clear was for the development of a system to build a union catalog for Ohio's research libraries through a regional library center and whose first major project would be a shared-cataloging activity. In retrospect, this paper provides a clue to the role that OCLC (which would several times change its name but fortunately not its initials) would play in future networking, as it is now called. He announced that "[a]t the present time, OCLC is not pursuing the mission of furnishing computation, but it has gently opened the door for access to this mission and will actively develop in this direction should the demand arise" (p. 82). Here, too, is set forth Kilgour's well-known philosophy for the making of personalized libraries for each user and his concept of OCLC as a node in the network of computerized textual material that would bring bibliographic resources close to each user (pp. 83-85).

In 1968, Kilgour also commented on the lack of techniques to organize a huge file of bibliographic entities and to provide the various indexes required for the envisioned system; also, terminals efficient enough for editing cataloging copy were nonexistent (p. 87). But new technology would overcome these problems and in October 1971, OCLC member libraries began to input new bibliographic data from their own terminals into the system (Long, p. 169). (The authors were privileged to be in Columbus to see OCLC in its first week of operation at this level.) "By February 1972, a system which permitted the smooth interaction of both member-created records and those from the Library of Congress was active for all but a very small percentage of Library of Congress records" (p. 169). By 1973, the OCLC cataloging system was servicing more than ninety terminals (seventy-five of which were in Ohio) while concluding cooperative agreements outside of Ohio with NELINET (New England Library Network), FAUL (Five Association University Libraries), PRLC (Pittsburgh Regional Library Center), PALINET (Pennsylvania Automated Library Network), and CCLC (Cooperative College Library Center). Installation of terminals in those libraries would double the number of operating online terminals (p. 169).

The extension of OCLC services to other groups throughout the nation grew, as did the database which eventually received data from hundreds of other libraries. OCLC produced (and still produces) many catalog cards—as well as many other products. Today, it would seem odd to hear a major paper at the Clinic describing the urgent needs of the Library of Congress Card Division. Yet Stephen Salmon (1970) presented such a paper at the 1969 Clinic for, at that time, LC was distributing over 110 million cards annually or about 1,000 cards every minute (p. 68). Today the needs of libraries still using cards are much more frequently met by the bibliographic utilities than by LC.

Although criticizing the bibliographic utilities for prolonging the card catalog's existence, Michael Gorman (1980) saw the producing of catalog cards by the utilities as a "transition" stage in library automation that, at the very least, had created "massive, centralized machine-readable data bases and individual machine-readable records" (p. 50). As early as the 1973 Clinic, Estelle Brodman (1974) heralded OCLC as one of the three most successful networks in librarianship. (The other two were LC's sale of catalog cards and NLM's MEDLINE.) This success, she believed, came because, in each of these three cases, the librarian could take standardized material and modify it in any way he or she wished (p. 21). When the 1976 Clinic attacked the problem of "Economics of Library Automation," Frederick G. Kilgour (1977) cited OCLC as an example of the computer's enhancing the productivity of library staff and reducing the need for complex codes such as the *Anglo-American Cataloging Rules* (pp. 6-8). Contrary to what might earlier have been expected, Kilgour thought that "[w]ith networks the national library will be made up of the nation's libraries, not any particular library" (p. 7).

Users and Online Catalogs

By the 1972 Clinic, I. A. Warheit (1973) was noting that "[l]ibrary technology is moving in the direction of on-line interactive processing" (p. 20). The central advantage was in bringing the information directly to the user so that he/she would not have to spend time and energy going to various sources (p. 16). The successful use of an online cataloging *system* was reported at that conference (Miller & Hodges, 1973). BALLOTS, the online interactive library automation system that would support the acquisitions and cataloging functions of the Stanford University Libraries' technical processing operations, was also described (Epstein & Veaner, 1975).

As it became apparent that many traditional manual catalogs would eventually give way to online computerized catalogs, librarians became concerned about how that new catalog would look. Would it mimic the traditional catalog, or would it become a very different catalog? How would it affect library catalog users, a long neglected group? At the Yale University Library, a research project carried out from 1967-69 had as its goal to "learn what a future catalog should be by studying, quantitatively, what our library patrons are trying, successfully or otherwise, to get out of our present catalog" (Lipetz, 1970, p. 43). Preliminary results of the study were reported at the 1969 Clinic.

The Yale study was performed with card catalog users, and R. G. Braithwaite (1979) reminded those attending the 1978 Clinic (as others before and after him did) that making an automated catalog did not mean automating the card catalog (p. 63). Getting rid of card catalog limitations, however, would introduce new computer-dependent limitations. In addition to technical problems, Braithwaite cautioned that there were problems because of the expectations of users who hold firmly to the belief that "the computer is a god and that its priests can do everything in no time at all" (p. 64).

In the Clinics of the late 1970s and early 1980s, it became more and more evident that more speakers were telling us that technology alone would not be enough to make online systems user friendly. Reference databases had presumed intermediaries would intervene (at least in the early stages) between users and databases, but from the beginning online catalogs were expected to be direct access tools for library users. These new means of service required not only an understanding of the capabilities and limitations but also an understanding of users for whom the systems were to be designed. Ward Shaw (1981) pointed out how little was really known about the design of public access systems. How to construct systems where hardware, software, people, and data would interact effectively would be a prime concern as online catalogs became realities. Principles of design of catalogs that would consider how fast a system should respond; how much data, and in what detail the data should be presented; how much should users become involved in or control the interactive process; and when should the system quit, were essentially new problems for the librarians to consider (pp. 3-4). All of this was complicated by the fact that librarians, in the words of Shaw, "don't know exactly how to produce informed (or satisfied) users; that is to say, we do not understand the mechanisms by which uninformed users are transformed into informed users" (p. 5). There was no escaping the need for a clear identification of goals and procedures to define the system, always within the context of a changing environment as new components became available (Avner, 1981, p. 18).

Studies to tell us more about users might be a new role for the library networks and, indeed, they accepted that role. Christine Borgman and Neal Kaske (1981) reported on a study to determine the number of terminals required for the Dallas Public Library catalog. While Dallas was the site of the study, the analysis of the collected data was done by computers at OCLC. The researchers did not attempt to extrapolate Dallas data to other libraries, but the results of this study did seem to confirm that usage of an online catalog was not the same as that of a card catalog (p. 35).

Studies at OCLC reported by Mary Ellen Jacob and Neal Kaske (1983) noted the connection between management information systems and online catalogs, which when used together, could, in the words of the reporters, "create an atmosphere for 'user studies deluxe!' The record of use that it is possible to create from an online catalog provides a library's management information system with accurate data on how the catalogs are being used by patrons" (p. 118). Through use of questionnaires, focused-group interviews, and transaction logs and activity reports, these studies revealed a number of barriers which prevent library patrons from effectively (or ever!) using library catalogs. The barriers included factors concerning the computers themselves, the system's language, and the bibliographic information (p. 119).

The design of interactive computer systems must consider more than bibliographic data as we have already intimated—how to manipulate the keyboard and how to deal with the information displayed must also become considerations. The content of the Clinics did not neglect these human considerations. Gregg Vanderheiden (1981) reported on methods to enhance, modify, and design terminals for access by the visually, physically, and cognitively disabled. W. David Penniman (1982) and his colleagues showed OCLC's concern in this area by discussing trends in input/output devices to improve displays and provide better access. While they saw room for much improvement, they felt the human factor would be the key to major improvements (p. 72). By the time an entire conference was devoted to *Human Aspects of Library Automation* (D. Shaw, 1986), four of the ten papers dealt specifically with the concern for library users' access to automated catalogs. Anne Gilliland (1986) discussed user reactions to online catalogs and indicated managerial decisions that should be considered regarding physical and intellectual aspects of online catalog access while Susan Roman (1986) delineated means that could be employed to enhance online catalog services for children and youth. Leslie Edmonds (1986) considered the needs for hardware and software designs for online systems that would serve physically disabled persons, the elderly, and non-English speakers, and Mark W. Arends (1986) gave helpful advice for designing instructional brochures for online catalog users.

The interest in human/computer interface continued in the following year with an exploration of user friendly as it applied to online catalogs and other library tools. Ward Shaw (1987) defined the goal of user friendliness as being "to provide a powerful, flexible, informative way for users to drive and control the system to their various ends" (p. 13). No easy task, according to Emily Fayen (1987), for users of online systems where an effective user interface "has to deal with complex record structures, data types, and files" and must provide "a comfortable tool for the occasional library user as well as the experienced staff member" (p. 55). Like others, she called for further research to build on that already completed in online catalog studies. Specifically, online catalog makers needed to be able to benefit from research that gave clues about the following questions: In what order should bibliographic records appear and should that order be the same for staff and public users? What fields of data should appear for each bibliographic record? What techniques work best to help users whose search strategy nets too few or too many results? How can one determine whether no hits for a search represents system failure or database failure? What is the best way to explain the basic idea of retrieval of sets to library users (pp. 58-59)?

At the same conference, Gary Golden (1987) extolled the use of microcomputers as public terminals at the University of Illinois at Urbana-Champaign Library. Interface programs in the microcomputers translated the natural language queries of users into the commands of the system, setting up interactions between the micro and the mainframe computer and making for a more user-friendly system.

As the progress of the catalog through the last quarter of the century is charted, some interesting observations come to mind. The familiar format of the catalog has undergone several major changes. Old ways have been discarded only to be rediscovered in new settings with new technologies. Catalogs have moved from being tools describing the collection of one library to tools describing the collections of a network or networks. Cataloging has become a truly shared effort—perhaps too much so—fostering a uniformity in many cases when it need not have been necessary to do so. There is a sameness in today's bibliographic data that speaks to the failure of extending the bibliographic content of the traditional record to being even more useful freed from the constraints of the 3 x 5 card. One frequently repeated message of Kilgour's was forgotten: that standardization and uniformity need not be the same. Users have gone from passive receivers in one location to interactive participants with catalogs available in a variety of locations. Users have managed to adapt to the new format, but their expectations have also increased. They thought all along that the computer would bring them a total bibliographic apparatus.

The 1980s became the decade of the online public access catalog (OPAC) but had we too soon become complacent with it? At the 1987 Clinic, Charles R. Hildreth (1989) summarized the development of the first generation online catalogs as tools consisting of derived key input, exact term or phrase matching, and lacking subject access. There was little in the way of online user assistance; no authority based searching with cross references was available; and meaningful browsing facilities were lacking, leading to criticisms that OPACs were actually inferior to the traditional library catalog. To Hildreth, in today's second-generation online catalogs, the traditional well-structured library catalog has joined with the power and flexibility of conventional information retrieval systems. Cross references are joined with information retrieval keyword and Boolean searching approaches. Searches can be restricted to specified record fields, character masking, and/or right hand truncation. Limits can be made by date, language, and place of publication. Records can be viewed and printed in a number of different display formats. Still, Hildreth finds much room for improvement. Even the second generation OPACs place the burden on the user to reformulate and reenter searches until satisfied. There are too many failed searches, too much confusion and frustration for users during the search process, unfamiliarity with or ignorance of the subject indexing, misunderstanding and confusion about fundamentally different approaches to retrieval and search methods, and missed opportunities to retrieve relevant materials. There is no automatic assistance with alternative search strategies and no leads from successful free-text search terms to corresponding subject heading or class numbers assigned to a broader range of related materials. The resulting bibliographic displays don't provide sufficient information to enable the user to judge the usefulness of documents, and there is no ranking of citations in regard to relevance to the user's search criteria. Hildreth concedes that too much focus has been placed on system performance factors with human searchers having been left out of improvements, and that there are still many technical, economic, and human problems to be solved before OPACs can be viewed as information retrieval systems and not just mechanized card catalogs. While some OPACs have the tools to integrate periodical and citation indexes into their catalogs, they cannot meet the price demanded by commercial suppliers of indexes and book review files. Yet Hildreth remains optimistic even though he sees current online catalog development as having slowed to a "snail's pace." The obstacles to the one stop, self service, information access and delivery station will be overcome in time. "The OPAC," he concludes, "has truly created an avalanche of possibilities and unleashed our imagination" (p. 32).

It appears, then, that this 25th conference is yet another link (albeit interface) in the journey begun in 1963 to see how catalogs have been affected by automation. The 25th conference includes two sessions relating to catalogs. As systems continue to develop, we expect that catalogs will be a continuing topic.

Technical Services: New Roles, New Organizational Patterns

Automation in libraries brought new roles, new organizational patterns, new requirements for technical services librarians. All these were parts of the Clinics. Librarians became involved in complex negotiations for contracts between libraries and other organizations such as regional networks for bibliographic utility services. Governance structures became important considerations, too (Evans, 1978). Librarians who developed databases found themselves in new roles as vendors when they shared the database with others (Upham & Wilcox, 1978).

Michael Gorman (1980) saw the integration of some processes and the increased role of cooperation and resource sharing as eliminating the need for division between public and technical services such as had existed in the past. Separate sections for handling special types of materials became unnecessary as organization by function rather than type of material became the norm. Nonprofessionals would perform much of the work while professionals would become policy makers and, to a limited extent, supervisors. All of this would result in a rethinking of the role of professional librarians.

When competencies for the new technologies were considered for the 1983 Clinic, Kathryn Luther Henderson (1984), from a review of the literature, a survey of individuals working in technical services positions, and an analysis of position announcements identified the competencies for technical services personnel as falling into two broad categories: (1) general, technical, and bibliographic; and (2) management, supervisory, and communicative (p. 13). Persons with analytical minds, problem solvers, decision makers and leaders would be a necessity. Such persons would need to be inquisitive, curious, imaginative, and creative. The technical services required adaptable and flexible persons amenable to change—as well as dreamers that envision new and better tools as the means to better service for users. In all of this, they should not lose sight of the purpose of their work (p. 36)—a point stressed at the first Clinic.

CONCLUSION

In the span of twenty-five years, many changes have occurred not only in technology but in human perceptions of library automation. Automation is a fact. Mistakes have been made and successes achieved. Some dreams have become realities while others faded upon rude awakenings. Some systems have been celebrated at their conception but secretly aborted and never heard of again. (Praise to the 1978 Clinic that was brave enough to admit to and record problems and failures! [Lancaster, 1979]) Unit record equipment, mainframes, minis, and micros have been a part of the progress and failure. Going it alone changed when the new technology could not be harnassed and managed on one's own. Acquisitions and serials management envisioned as prime candidates for automated activities in 1963 did not flourish in the way that cataloging did. The lowly catalog, doomed to extinction by earlier prophets, now seemed destined to become the more comprehensive tool that librarians in the mid-nineteenth century envisioned but gave up when catalog formats were more cumbersome. A new era of sharing emerged. Librarians took on new roles and realigned old competencies. Renewed emphasis has been given to library users and systems responsive to their needs. Yet there is much left to be done.

All of this has affected the functions, the organization of, and the people associated with technical services as have been mirrored by the content of the past twenty-four Clinics. Despite the fact that many factors influence the content of an individual conference, these Clinics, as a whole, have in many ways been ahead of their time in topics covered in their presentations and provided rewarding, educational experiences.

In a scenario in the 1979 Clinic, F. Wilfrid Lancaster, Laura S. Drasgow, and Ellen B. Marks saw a dwindling away of the technical services in general by the twenty-first century (Lancaster, 1980, pp. 179-81). That will be for the speakers at the fiftieth anniversary in 2013 to comment upon. The authors, who have together accumulated and enjoyed over six decades of work in the technical services or teaching others about those services, rather hope that scenario will not totally be played out.

REFERENCES

Adkinson, B. W. (1964). Trends in library applications of data processing. In H. Goldhor (Ed.), *Proceedings of the 1963 clinic on library applications of data processing* (Papers presented at the 1st Annual Clinic, 28 April-1 May 1963) (pp. 1-8). Urbana-Champaign: University of Illinois, Graduate School of Library Science.

Auld, L., & Baker, R. (1972). LOLITA: An online book order and fund accounting system. In F. W. Lancaster (Ed.), *Applications of online computers to library problems* (Papers presented at the 9th Annual Clinic on Library Applications of Data Processing, 30 April-3 May 1972) (pp. 29-53). Urbana-Champaign: University of Illinois, Graduate School of Library Science.

Arends, M. W. (1986). Designing effective instructional brochures for online catalogs. In D. Shaw (Ed.), *Human aspects of library automation: Helping staff and patrons cope* (Papers presented at the 22nd Annual Clinic on Library Applications of Data Processing, 14-16 April 1985) (pp. 108-16). Urbana-Champaign: University of Illinois, Graduate School of Library and Information Science.

Avram, H. D. (1971). The evolving MARC system: The concept of a data utility. In K. L. Henderson (Ed.), *MARC uses and users* (Papers presented at the 8th Annual Clinic on Library Applications of Data Processing, 26-29 April 1970) (pp. 1-26). Urbana-Champaign: University of Illinois, Graduate School of Library Science.

Avner, A., & Friedman, H. G., Jr. (1981). Interacting with computer users: Design considerations. In J. L. Divilbiss (Ed.), *Public access to library automation* (Papers presented at the 17th Annual Clinic on Library Applications of Data Processing, 20-23 April 1980) (pp. 8-19). Urbana-Champaign: University of Illinois, Graduate School of Library and Information Science.

Black, D. V. (1969). Library information system time-sharing on a large, general-purpose computer. In D. E. Carroll (Ed.), *Proceedings of the 1968 clinic on library applications of data processing* (Papers presented at the 6th Annual Clinic) (pp. 139-54). Urbana-Champaign: University of Illinois, Graduate School of Library Science.

Borgman, C. L., & Kaske, N. K. (1981). Determining the number of terminals required for an on-line catalog through queueing analysis of catalog traffic data. In J. L. Divilbiss (Ed.), *Public access to library automation* (Papers presented at the 17th Annual Clinic on Library Applications of Data Processing, 20-23 April 1980) (pp. 20-36). Urbana-Champaign: University of Illinois, Graduate School of Library and Information Science.

Braithwaite, R. J. (1979). Automation of the catalog: The transition from cards to computers. In F. W. Lancaster (Ed.), *Problems and failures in library automation* (Papers presented at the 15th Annual Clinic on Library Applications of Data Processing, 23-26 April 1978) (pp. 60-66). Urbana-Champaign: University of Illinois, Graduate School of Library Science.

Brezgis, R. (1966). The ONLUP bibliographic control system: An evaluation. In F. B. Jenkins (Ed.), *Proceedings of the 1965 clinic on library applications of data processing* (Papers presented at the 3rd Annual Clinic, 25-28 April 1965) (pp. 112-40). Urbana-Champaign: University of Illinois, Graduate School of Library Science.

Brodman, E. (1974). Backing into network operations. In F. W. Lancaster (Ed.), *Networking and other forms of cooperation* (Papers presented at the 10th Annual Clinic on Library Applications of Data Processing, 29 April-2 May 1973) (pp. 9-23). Urbana-Champaign: University of Illinois, Graduate School of Library Science.

Brody, A. (1966). Bro-Dart Industries' experience with electronic data processing. In F. B. Jenkins (Ed.), *Proceedings of the 1965 clinic on library applications of data processing* (Papers presented at the 3rd Annual Clinic, 25-28 April 1965) (pp. 65-78). Urbana-Champaign: University of Illinois, Graduate School of Library Science.

Brudvig, G. L. (1974). The development of a minicomputer system for the University of Minnesota Bio-Medical Library. In F. W. Lancaster (Ed.), *Applications of minicomputers to library and related problems* (Papers presented at the 11th Annual Clinic on Library Applications of Data Processing, 28 April-1 May 1974) (pp. 170-80). Urbana-Champaign: University of Illinois, Graduate School of Library Science.

Burgess, T. K. (1970). Criteria for design of an on-line acquisitions system at Washington State University Library. In D. E. Carroll (Ed.), *Proceedings of the 1969 clinic on library applications of data processing* (Papers presented at the 7th Annual Clinic, 27-30 April 1969) (pp. 50-66). Urbana-Champaign: University of Illinois, Graduate School of Library Science.

Burke, J. (1986). Automation planning and implementation: Library and vendor responsibilities. In D. Shaw (Ed.), *Human aspects of library automation: Helping staff and patrons cope* (Papers presented at the 22nd Annual Clinic on Library Applications of Data Processing, 14-16 April 1985) (pp. 46-58). Urbana-Champaign: University of Illinois, Graduate School of Library and Information Science.

Burns, L. R. (1964). Automation in the public libraries of Lake County, Indiana. In H. Goldhor (Ed.), *Proceedings of the 1963 clinic on library applications of data processing* (Papers presented at the 1st Annual Clinic, 28 April-1 May 1963) (pp. 9-17). Urbana-Champaign: University of Illinois, Graduate School of Library Science.

Camp, J. A.; Agnew, G.; Landram, C.; Richards, J.; & Shelton, J. (1987). Survey of online systems in U. S. academic libraries. *College and Research Libraries, 48*(4), 339-350.

Coward, R. E. (1971). The BNB MARC project. In K. L. Henderson (Ed.), *Proceedings of the 1970 clinic on library applications of data processing* (Papers presented at the 8th Annual Clinic, 26-29 April 1970) (pp. 36-52). Urbana-Champaign: University of Illinois, Graduate School of Library Science.

Cox, C. R. (1965). Review of 1963 clinic on library applications of data processing. *College and Research Libraries, 26*(5), 409.

Culbertson, D. S.; Voigt, M. J.; & Cox, J. R. (1963). The costs of data processing in university libraries. *College and Research Libraries, 24* (November), 487-495.

Curley, W. W. (1966). The data processing program in operation at the Suffolk Cooperative Library System, Patchogne, New York. In F. B. Jenkins (Ed.), *Proceedings of the 1965 clinic on library applications of data processing* (Papers presented at the 3rd Annual Clinic, 25-28 April 1965) (pp. 15-42). Urbana-Champaign: University of Illinois, Graduate School of Library Science.

Curran, A. J. (1970). Library networks: Cataloging and bibliographic aspects. In D. E. Carroll (Ed.), *Proceedings of the 1969 clinic on library applications of data processing* (Papers presented at the 7th Annual Clinic, 27-30 April 1969) (pp. 31-41). Urbana-Champaign: University of Illinois, Graduate School of Library Science.

Derthicke, J., & Moran, B. B. (1986). Serials agent selection in ARL libraries. In M. Tuttle and J. G. Cook (Eds.), *Advances in serial management, Vol. 1* (p. 14). Greenwich, CT: JAI Press.

Dodendorf, M. S. (1966). 870 Document Writing System of International Business Machine Corporation in the library section, Los Angeles city schools. In F. B. Jenkins (Ed.), *Proceedings of the 1965 clinic on library applications of data processing* (Papers presented at the 3rd Annual Clinic, 25-28 April 1965) (pp. 43-64). Urbana-Champaign: University of Illinois, Graduate School of Library Science.

Edelman, H., & Muller, K. (1987). A new look at the library market. *Publishers' Weekly, 231* (May 27), 34.

Edmonds, L. (1986). Online services and specialized clienteles: Handicapped and other populations. In D. Shaw (Ed.), *Human aspects of library automation: Helping staff and patrons cope* (Papers presented at the 22nd Annual Clinic on Library Applications of Data Processing, 14-16 April 1985) (pp. 101-07). Urbana-Champaign: University of Illinois, Graduate School of Library and Information Science.

Epstein, A. H., & Veaner, A. B. (1973). A user's view of BALLOTS. In F. W. Lancaster (Ed.), *Applications of on-line computers to library problems* (Papers presented at the 9th Annual Clinic on Library Applications of Data Processing, 30 April-3 May 1972) (pp. 109-37). Urbana-Champaign: University of Illinois, Graduate School of Library Science.

Evans, G. T. (1978). Regional network contracts with libraries for OCLC services. In J. L. Divilbiss (Ed.), *Negotiating for computer services* (Papers presented at the 15th Annual Clinic on Library Applications of Data Processing, 24-27 April 1977) (pp. 9-22). Urbana-Champaign: University of Illinois, Graduate School of Library and Information Science.

Fasana, P. J. (1963). Automating cataloging functions in conventional libraries. *Library Resources and Technical Services,* 7(4), 350-365.

Fayen, E. G. (1987). User interfaces for online library catalogs. In F. W. Lancaster (Ed.), *What is user friendly?* (Papers presented at the 23rd Annual Clinic on Library Applications of Data Processing, 20-22 April 1986) (pp. 52-60). Urbana-Champaign: University of Illinois, Graduate School of Library and Information Science.

Fayollat, J. (1973). On-line serials control at UCLA. In F. W. Lancaster (Ed.), *Applications of on-line computers to library problems* (Papers presented at the 9th Annual Clinic on Library Applications of Data Processing, 30 April-3 May 1972) (pp. 69-81). Urbana-Champaign: University of Illinois, Graduate School of Library Science.

Freedman, M. J. (1977). The economics of book catalog production. In J. L. Divilbiss (Ed.), *The economics of library automation* (Papers presented at the 13th Annual Clinic on Library Applications of Data Processing, 25-28 April 1976) (pp. 107-21). Urbana-Champaign: University of Illinois, Graduate School of Library Science.

Gilliland, A. (1986). Online catalogs and library users. In D. Shaw (Ed.), *Human aspects of library automation: Helping staff and patrons cope* (Papers presented at the 22nd Annual Clinic on Library Applications of Data Processing, 14-16 April 1985) (pp. 77-93). Urbana-Champaign: University of Illinois, Graduate School of Library and Information Science.

Golden, G. A. (1987). Taming the unfriendly system: Microcomputers as patron terminals to access an online catalog. In F. W. Lancaster (Ed.), *What is user friendly?* (Papers presented at the 23rd Annual Clinic on Library Applications of Data Processing,

20-22 April 1986) (pp. 61-79). Urbana-Champaign: University of Illinois, Graduate School of Library and Information Science.

Goldhor, H. (1964). Foreword. In H. Goldhor (Ed.), *Proceedings of the 1963 clinic on library applications of data processing* (Papers presented at the 1st Annual Clinic, 28 April-1 May 1963) (pp. ii-v). Urbana-Champaign: University of Illinois, Graduate School of Library Science.

Gorman, M. (1977). The economics of book catalog production. In J. L. Divilbiss (Ed.), *The economics of library automation* (Papers presented at the 13th Annual Clinic on Library Applications of Data Processing, 25-28 April 1976) (pp. 122-32). Urbana-Champaign: University of Illinois, Graduate School of Library Science.

Gorman, M. (1987). Linking the unlinkable. In F. W. Lancaster (Ed.), *What is user friendly?* (Papers presented at the 23rd Annual Clinic on Library Applications of Data Processing, 20-22 April 1986) (pp. 2-8). Urbana-Champaign: University of Illinois, Graduate School of Library and Information Science.

Gorman, M. (1980). Technical services in an automated library. In F. W. Lancaster (Ed.), *The role of the library in an electronic society* (Papers presented at the 16th Annual Clinic on Library Applications of Data Processing, 22-25 April 1979) (pp. 95-103). Urbana-Champaign: University of Illinois, Graduate School of Library Science.

Griffin, H. (1964). Electronic data processing applications to technical processing and circulation activities in a technical library. In H. Goldhor (Ed.), *Proceedings of the 1963 clinic on library applications of data processing* (Papers presented at the 1st Annual Clinic, 28 April-1 May 1963) (pp. 96-108). Urbana-Champaign: University of Illinois, Graduate School of Library Science.

Griffin, H. (1971). MARC users: A study of the distribution of MARC tapes and the subscribers to MARC. In K. L. Henderson (Ed.), *MARC uses and users* (Papers presented at the 8th Annual Clinic on Library Applications of Data Processing, 26-29 April 1970) (pp. 27-35). Urbana-Champaign: University of Illinois, Graduate School of Library Science.

Griffin, M. (1964). IBM advanced systems development library in transition. In H. Goldhor (Ed.), *Proceedings of the 1963 clinic on library applications of data processing* (Papers presented at the 1st Annual Clinic, 28 April-1 May 1963) (pp. 79-95). Urbana-Champaign: University of Illinois, Graduate School of Library Science.

Henderson, J. D. (1964). The book catalogs of the Los Angeles County Public Library. In H. Goldhor (Ed.), *Proceedings of the 1963 clinic on library applications of data processing* (Papers presented at the 1st Annual Clinic, 28 April-1 May 1963) (pp. 18-36). Urbana-Champaign: University of Illinois, Graduate School of Library Science.

Henderson, K. L. (1971). Introduction. In K. L. Henderson (Ed.), *MARC uses and users* (Papers presented at the 8th Annual Clinic on Library Applications of Data Processing, 26-29 April 1970) (pp. ix-x). Urbana-Champaign: University of Illinois, Graduate School of Library Science.

Henderson, K. L. (1983). The new technology and competencies for "the most typical of the activities of libraries": Technical services. In L. C. Smith (Ed.), *Professional competencies—Technology and the librarian* (Papers presented at the 20th Annual Clinic on Library Applications of Data Processing, 24-26 April 1983) (pp. 12-42). Urbana-Champaign: University of Illinois, Graduate School of Library and Information Science.

Hildreth, C. R. (1988). Extending the access and reference service capabilities of the online public access catalog. In L. C. Smith (Ed.), *Questions and answers: Strategies for using the electronic reference collection* (Papers presented at the 24th Annual Clinic on Library Applications of Data Processing, 5-7 April 1987) (pp. 14-33). Urbana-Champaign: University of Illinois, Graduate School of Library and Information Science.

Jacob, M. E., & Kaske, N. K. (1983). Management information systems in a network environment. In F. W. Lancaster (Ed.), *Library automation as a source of management information* (Papers presented at the 19th Annual Clinic on Library Applications of Data Processing, 25-28 April 1982) (pp. 111-27). Urbana-Champaign: University of Illinois, Graduate School of Library and Information Science.

Kennedy, J. P. (1969). A local MARC project: The Georgia Tech Library. In D. E. Carroll (Ed.), *Proceedings of the 1968 clinic on library applications of data processing* (Papers presented at the 6th Annual Clinic, 5-8 May 1968) (pp. 199-215). Urbana-Champaign: University of Illinois, Graduate School of Library Science.

Kilgour, F. G. (1965). Development of computerization of card catalogs in medical and scientific libraries. In H. Goldhor (Ed.), *Proceedings of the 1964 clinic on library applications of data processing* (Papers presented at the 2nd Annual Clinic, 26-29 April 1964) (pp. 25-35). Urbana-Champaign: University of Illinois, Graduate School of Library Science.

Kilgour, F. G. (1977). The economics of library computerization. In J. L. Divilbiss (Ed.), *The economics of library automation* (Papers presented at the 13th Annual Clinic on Library Applications of Data Processing, 25-28 April 1976) (pp. 3-9). Urbana-Champaign: University of Illinois, Graduate School of Library Science.

Kilgour, F. G. (1969). Initial system design for the Ohio Collect[sic] Library Center: A case history. In D. E. Carroll (Ed.), *Proceedings of the 1968 clinic on library applications of data processing* (Papers presented at the 6th Annual Clinic, 5-8 May 1968) (pp. 79-88). Urbana-Champaign: University of Illinois, Graduate School of Library Science.

Kilgour, F. G. (1971). Summary of MARC. In K. L. Henderson (Ed.), *MARC uses and users* (Papers presented at the 8th Annual Clinic on Library Applications of Data Processing, 26-29 April 1970) (pp. 105-10). Urbana-Champaign: University of Illinois, Graduate School of Library Science.

King, G.; Edmundson, H. P.; Flood, M. M.; Kochen, M.; Libby, R. L.; Swanson, D. R.; & Wylly, A. (1963). *Automation and the Library of Congress.* Washington, DC: Library of Congress.

Kozlow, R. D. (1969). Genesis for a serials list. In D. E. Carroll (Ed.), *Proceedings of the 1968 clinic on library applications of data processing* (Papers presented at the 6th Annual Clinic, 5-8 May 1968) (pp. 216-35). Urbana-Champaign: University of Illinois, Graduate School of Library Science.

Lancaster, F. W.; Drasgow, L. S.; & Marks, E. B. (1980). The role of the library in an electronic society. In F. W. Lancaster (Ed.), *The role of the library in an electronic society* (Papers presented at the 16th Annual Clinic on Library Applications of Data Processing, 22-25 April 1979) (pp. 162-91). Urbana-Champaign: University of Illinois, Graduate School of Library Science.

Long, P. (1974). OCLV: From concept to functioning network. In F. W. Lancaster (Ed.), *Networking and other forms of cooperation* (Papers presented at the 10th Annual Clinic on Library Applications of Data Processing, 29 April-2 May 1973) (pp. 165-70). Urbana-Champaign: University of Illinois, Graduate School of Library Science.

Lipetz, B. (1970). A quantitative study of catalog use. In D. E. Carroll (Ed.), *Proceedings of the 1969 clinic on library applications of data processing* (Papers presented at the 7th Annual Clinic, 27-30 April 1969) (pp. 42-49). Urbana-Champaign: University of Illinois, Graduate School of Library Science.

Malinconico, S. M. (1977). The economics of computer output media. In J. L. Divilbiss (Ed.), *The economics of library automation* (Papers presented at the 13th Annual Clinic on Library Applications of Data Processing, 25-28 April 1976) (pp. 145-62). Urbana-Champaign: University of Illinois, Graduate School of Library Science.

Markuson, B. E. (1968). Aspects of automation viewed from the Library of Congress. In D. E. Carroll (Ed.), *Proceedings of the 1967 clinic on library applications of data processing* (Papers presented at the 5th Annual Clinic, 30 April-3 May 1967) (pp. 98-129). Urbana-Champaign: University of Illinois, Graduate School of Library Science.

Markuson, B. E. (Ed.). (1964). *Libraries and automation.* (Papers presented at the Conference on Libraries and Automation held at Airlie Foundation, Warrenton, Virginia, 26-30 May 1963). Washington, DC: Library of Congress.

McGee, R. S., & Miller, R. C. (1971). MARC utilization in the University of Chicago

Library Bibliographic Data Processing System. In K. L. Henderson (Ed.), *MARC uses and users* (Papers presented at the 8th Annual Clinic on Library Applications of Data Processing, 26-29 April 1970) (pp. 78-97). Urbana-Champaign: University of Illinois, Graduate School of Library Science.

Miller, E. W., & Hodges, B. J. (1973). Shawnee Mission's on-line cataloging system: The first two years. In F. W. Lancaster (Ed.), *Applications of on-line computers to library problems* (Papers presented at the 9th Annual Clinic on Library Applications of Data Processing, 30 April-3 May 1972) (pp. 94-108). Urbana-Champaign: University of Illinois, Graduate School of Library Science.

The new international yearbook for the year 1963. (1964). New York: Funk and Wagnalls.

Parker, R. H. (1964). Development of automatic systems at the University of Missouri Library. In H. Goldhor (Ed.), *Proceedings of the 1963 clinic on library applications of data processing* (Papers presented at the 1st Annual Clinic, 28 April-1 May 1963) (pp. 43-55). Urbana-Champaign: University of Illinois, Graduate School of Library Science.

Payne, C. T. (1967). An integrated computer-based bibliographic data system for a large university library: Problems and progress at the University of Chicago. In D. E. Carroll (Ed.), *Proceedings of the 1967 clinic on library applications of data processing* (Papers presented at the 5th Annual Clinic, 30 April-3 May 1967) (pp. 29-40). Urbana-Champaign: University of Illinois, Graduate School of Library Science.

Penniman, W. D.; Tuttle, H.; & Hickey, T. S. (1982). New technologies and opportunities regarding input/output devices. In L. C. Smith (Ed.), *New information technologies — New opportunities* (Papers presented at the 18th Annual Clinic on Library Applications of Data Processing, 26-29 April 1981) (pp. 60-73). Urbana-Champaign: University of Illinois, Graduate School of Library and Information Science.

Richmond, P. A. (1963). A short-title catalog made with IBM tabulating equipment. *Library Resources and Technical Services, 7*(1), 81-90.

Roman, S. (1986). Online catalogs and specialized clienteles: Children and youth. In D. Shaw (Ed.), *Human aspects of library automation: Helping staff and patrons cope* (Papers presented at the 22nd Annual Clinic on Library Applications of Data Processing, 14-16 April 1985) (pp. 94-100). Urbana-Champaign: University of Illinois, Graduate School of Library and Information Science.

Salmon, S. R. (1970). Development of the card-automated reproduction and distribution system (CARDS) at the Library of Congress. In D. E. Carroll (Ed.), *Proceedings of the 1969 clinic on library applications of data processing* (Papers presented at the 7th Annual Clinic, 27-30 April 1969) (pp. 98-113). Urbana-Champaign: University of Illinois, Graduate School of Library Science.

Sage, C. R. (1974). Utilization of the MARC II format for serials in an inter-university environment. In F. W. Lancaster (Ed.), *Networking and other forms of cooperation* (Papers presented at the 10th Annual Clinic on Library Applications of Data Processing, 29 April-2 May 1973) (pp. 24-31). Urbana-Champaign: University of Illinois, Graduate School of Library Science.

Salton, G. (1980). Toward a dynamic library. In F. W. Lancaster (Ed.), *The role of the library in an electronic society* (Papers presented at the 16th Annual Clinic on Library Applications of Data Processing, 22-25 April 1979) (pp. 60-81). Urbana-Champaign: University of Illinois, Graduate School of Library Science.

Schultheiss, L. A. (1964). Techniques of flow-charting. In H. Goldhor (Ed.), *Proceedings of the 1963 clinic on library applications of data processing* (Papers presented at the 1st Annual Clinic, 28 April-1 May 1963) (pp. 62-78). Urbana-Champaign: University of Illinois, Graduate School of Library Science.

Schultheiss, L. A.; Culbertson, D. S.; & Heiliger, E. M. (1962). *Advanced data processing in the university library.* New York: Scarecrow Press.

Segesta, J., & Hersberger, R. (1987). A quarter century of advanced data processing in the university library. *College and Research Libraries, 48*(5), 399-407.

Shaw, D. (Ed.). (1986). *Human aspects of library automation: Helping staff and patrons cope* (Papers presented at the 22nd Annual Clinic on Library Applications of Data

Processing, 14-16 April 1985). Urbana-Champaign: University of Illinois, Graduate School of Library and Information Science.

Shaw, W. (1981). Design principles for public access. In J. L. Divilbiss (Ed.), *Public access to library automation* (Papers presented at the 17th Annual Clinic on Library Applications of Data Processing, 20-23 April 1980) (pp. 2-7). Urbana-Champaign: University of Illinois, Graduate School of Library and Information Science.

Shaw, W. (1987). Aristotle meets Plato in the library catalog: Part 1. In F. W. Lancaster (Ed.), *What is user friendly?* (Papers presented at the 23rd Annual Clinic on Library Applications of Data Processing, 20-22 April 1986) (pp. 9-14). Urbana-Champaign: University of Illinois, Graduate School of Library and Information Science.

Stromberg, D. H. (1966). Computer applications to book catalogs and library systems. In H. Goldhor (Ed.), *Proceedings of the 1966 clinic on library applications of data processing* (Papers presented at the 4th Annual Clinic, 24-27 April 1966) (pp. 195-210). Urbana-Champaign: University of Illinois, Graduate School of Library Science.

Taine, S. I. (1964). Bibliographic data processing at the National Library of Medicine. In H. Goldhor (Ed.), *Proceedings of the 1963 clinic on library applications of data processing* (Papers presented at the 1st Annual Clinic, 28 April-1 May 1963) (pp. 109-24). Urbana-Champaign: University of Illinois, Graduate School of Library Science.

Tauber, M. F. (1964). Technical services in 1963. *Library Resources & Technical Services, 8*(2), 101-111.

Thomson, J. W., & Muller, R. H. (1969). The computer-based book order system at the University of Michigan Library: A review and evaluation. In D. E. Carroll (Ed.), *Proceedings of the 1968 clinic on library applications of data processing* (Papers presented at the 6th Annual Clinic, 5-8 May 1968) (pp. 54-78). Urbana-Champaign: University of Illinois, Graduate School of Library Science.

University of Illinois, Graduate School of Library and Information Science. (1962, September 11). *Minutes of the faculty meeting.* Urbana-Champaign: University of Illinois, GSLIS.

University of Illinois, Graduate School of Library and Information Science. (1963, March 28). *Minutes of the faculty meeting.* Urbana-Champaign: University of Illinois, GSLIS.

Upham, L., & Wilcox, A. (1978). Negotiating for data base sharing. In J. L. Divilbiss (Ed.), *Negotiating for computer services* (Papers presented at the 14th Annual Clinic on Library Applications of Data Processing, 24-27 April 1977) (pp. 95-103). Urbana-Champaign: University of Illinois, Graduate School of Library Science.

Vanderheiden, G. (1981). Modifying and designing computer terminals to allow access by handicapped individuals. In J. L. Divilbiss (Ed.), *Public access to library automation* (Papers presented at the 17th Annual Clinic on Library Applications of Data Processing, 20-23 April 1980) (pp. 99-116). Urbana-Champaign: University of Illinois, Graduate School of Library and Information Science.

Warheit, I. A. (1972). On-line interactive systems in libraries now and in the future. In F. W. Lancaster (Ed.), *Applications of on-line computers to library problems* (Papers presented at the 9th Annual Clinic on Library Applications of Data Processing, 30 April-3 May 1972) (pp. 3-21). Urbana-Champaign: University of Illinois, Graduate School of Library Science.

Year, 1964 Encyclopedia news annual, Events of the year 1963. (1963). New York: Year, Inc.

Zuckerman, R. A. (1968). Computerized book catalogs and their effects on integrated library data processing: Research and progress at the Los Angeles County Public Library. In D. E. Carroll (Ed.), *Proceedings of the 1967 clinic on library applications of data processing* (Papers presented at the 5th Annual Clinic, 30 April-3 May 1967) (pp. 70-89). Urbana-Champaign: University of Illinois, Graduate School of Library Science.

RICHARD RUBIN

Assistant Professor
School of Library Science
Kent State University
Kent, Ohio

The Management of Automation: A Review of the Proceedings of the Data Processing Clinics

INTRODUCTION

The topic for this presentation is the Management of Library Automation as viewed through the twenty-five years of data processing clinic proceedings. In a way, it is a disconcerting topic, because it generates ambivalence: have librarians managed automation, or has it managed librarians? The author's experience suggests that the introduction of new technology stimulates in employees either cynicism or a powerful existential angst. Predictably, the managerial pose that is struck when employees express trepidation concerning new technology is that they (the employees) must adapt; that the key to dealing with automation is (the employees') open-mindedness and flexibility; and that it is their (the employees') defects—mental, emotional, or physical—that threaten the success of automation.

It is not surprising, then, that much of the current management literature, including a recent edition of the proceedings (Shaw, 1985), concentrates on why employees fear and resist technology, and how employers might dispel their misgivings.

To manage is to control, and the library literature on managing automation is one part the literature of controlling the machine and one part the literature of controlling the employee. This latter concern simply recognizes that, to a large extent, the machine has profoundly affected how one manages oneself. Automation changes the tasks and responsibilities of one's job, redefines one's organizational and departmental roles, alters one's work climate, restructures one's fiscal environment, and has brought into the workplace new employees whose interests, skills, and language are, to many librarians, peculiar, even bizarre. In the final analysis, the literature of managing automation is divided into the literature of managing the machine and the literature of managing the people.

The balance of this paper is organized into three parts. Each part reviews in chronological order pertinent proceedings from the 1960s, 1970s, and 1980s.

PHASE I. FOCUS ON THE MACHINE

The Proceedings of the 1960s

Very early editions of the proceedings were not concerned with management issues per se; rather, they dealt with characteristics of the automated systems themselves. References to management were at best incidental.

One of the early proceedings (1967) did include two articles on the management of automation. The first was a case study of an automated system (Hage, 1967). It ostensibly covered such topics as consultants' reports, bidding, and staff involvement. In reality, it was a paean to the computer, a uniformly optimistic assessment. This optimism was natural and certainly not uncommon for the day. There was, however, a more discerning article in the same volume on "The Decision to Automate" (Chapin, 1967). Today, the decision whether to automate seems almost quaint, although it may still arise in some library backwaters. Now managers are more often concerned with which system to automate rather than whether to automate. But in Chapin's day (it seems as though this took place in the nineteenth century, rather than merely two decades ago!) the desirability of automation was a legitimate question. What factors did the manager consider in making this decision?

Although managers are seldom drawn to philosophical musings, Chapin did engage in one, more speculative reflection: perhaps, he mused, implementing automation would ultimately lead to the decline of reading and writing. While this was only a rhetorical foreboding, two more tangible problems affected the decision to automate: the costs of automation were uncertain, and the technology had significant deficiencies. (For example, scanners were having difficulty reading different type fonts.) These observations were both practical and central to the management of automation at this time. They focused on the potential liabilities of the machine itself.

Despite these misgivings, Chapin found many reasons for the

manager to consider automation in 1967. These reasons can be grouped into four categories:

1. *The need to cut costs.* The library was already experiencing the inflationary pressures that were to degrade the dollar for a decade, and there was no reason to assume that the costs of materials or labor were heading downward. Similarly, there was little reason to believe that library budget increases would offset these costs.
2. *Increased demands on the part of the users.* The patron was demanding better access to the literature. If the manager's goal was service, then something had to be done to improve it. User frustration was increasing.
3. *The expansion of publications.* The "information explosion" had arrived. Control over the literature of the sciences had become an especially imposing task.
4. *Understaffing.* Libraries had too few employees to provide the needed services. Automation might provide maximum efficiency for the already burdened library work force.

Certainly from a management perspective, Chapin was performing an important function: identifying forces both environmental and internal that affected productivity in the organization and dictated change. When he identified these forces, he provided a fundamental rationalization for the decision to automate.

Of course, identifying the need for a change does not in itself suggest how one changes judiciously. To this end, Chapin provided some general managerial advice to those who were considering the automated road not yet taken. His concerns included:

— How much of the library will be automated?

— Will the system function with more than books?

— Is the system adaptable to online use?

— What types of information will be provided by the system?

— Will the system yield cost information?

— How will the system be evaluated? By cost? By currency and accuracy? By ability to handle increased load? By acceptance of staff and users?

These were certainly reasonable questions for the library manager to consider, and their appearance in the 1967 proceedings confirms that the focus of the management of automation was on the system itself, not on the people. But the two articles in the 1967 proceedings must be considered anomalous. Management issues were not to take a prominent position for some years to come.

Other management articles did appear sporadically. For example, T. C. Dobb (1970) from the Simon Fraser Computer Centre in British Columbia wrote on the organization of data processing for the library from the perspective of the computer center. Perhaps his most salient observation was that, when it comes to automation, the organizational structure was not as important as the people selected to fill the positions within that structure. It was vaguely reminiscent of the battle in organizational theory between sociologist Max Weber, who emphasized rational bureaucratic structure, and management theorist Douglas MacGregor, who emphasized the importance of human motivation. Does a rationalized bureaucratic structure provide maximum productivity, or does the "right" employee provide the needed productivity regardless of the structure? On this issue, Dobb had an international inspiration. He crystallized his thoughts by modifying what he called an old Chinese proverb:

> If the wrong people are in the right structure, the right structure will work in the wrong way; but if the right people are in the wrong structure, the wrong structure will work in the right way. (p. 80)

Dobb does not say where he found the original version, nor what the original creator would have thought of this adaptation. But the message was clear: the people were critical.

The reason why articles on the management of automation were sporadic in the early years of the proceedings may have been a natural outgrowth of the incipient character of library automation. Drawing from the work of Henry Lucas, John Olsgaard (1985) from the University of South Carolina has suggested that the literature of computer systems, indeed the development of the discipline of computer systems itself, follows a linear progression: first are considerations primarily involving technological or physical issues (the machine); second come organizational considerations (the structure); and third are considerations related to organizational behavior (the people) (p. 20).

That the proceedings would reflect this linear progression is logical. The early proceedings concerned themselves with technological problems of the machine. These were the problems immediately confronting the library decision-maker. Discussion of the organizational structure and the management of people would have to wait until the manager had implemented automation and discovered the human problems that lay in ambush. A reservoir of managerial experience was necessary before it was possible to create substantive generalities in these areas.

PHASE II. ORGANIZATIONAL CONSIDERATIONS

Proceedings of 1976-78

The proceedings from 1976-78 reflect the evolution of management concerns from the technical aspects of the machine itself to the organizational issues affected by the implementation of automation. Three basic management topics were reviewed over these three years of the proceedings: the economics of automation, contract negotiations, and causes of failure in library automation. In each case, one common characteristic is manifest: time—time to gather economic data, time to fall victim to vendors, time to experience the agony of technological defeat. It should be noted, however, that even now the discussion focused on the effects of the machine on the organization rather than the effects employees had on the organization.

The 1976 proceedings was devoted to the economics of library automation. As might be predicted, part of the economic picture involved the ubiquitous assertion that computers could reduce labor costs. A stimulating article entitled "The Economics of Library Computerization" described the fiscal threats that had descended on the library, and proposed the use of "scientific economics" as a tool for assessing library automation (Kilgour, 1976). Kilgour observed that despite the fact that libraries were being managed well, they were in financial peril. This peril arose because costs were rising, including staff costs, and a substantial portion of library patrons were failing to get what they wanted from their libraries. Increasing costs and decreasing service is, of course, exactly what a manager does not want to hear. That automation provided perhaps the only means for ameliorating the situation seemed obvious. Among the labor-saving areas noted were:

— automation increases the amount of work done,

— computers can substitute for human effort,

— computers are faster than humans, and

— computers allow for automatic detection of error. (p. 6)

Kilgour further argued that economies of scale, especially in such areas as shared cataloging, were particularly appropriate for computer application. His closing statement was a clarion call to the reluctant library manager still percolating over the question: "Should I or shouldn't I automate?"

> It is all too clear from economic analysis that libraries have extremely serious problems to be solved. There is no way that society is going to support a 460 percent increase in financial support for an institution experiencing a 50-60 percent failure rate in service. Libraries are as efficient as other labor-intensive service industries, and it is impossible to see how any further increase in the efficiency of an already highly efficient operation can cope with such rocketing increases in costs. It is inevitable that a drastic change must occur in library operations; for the immediate future, the greatest desirable impact will come from computerized, on-line networking that provides not only labor-saving functions but also effective economies of scale. (p. 9)

It is a perplexing paragraph. How can libraries be called "highly efficient" yet be unable to provide satisfactory service 50 percent of the time? One detects overstatement and flattery in the claim of efficiency. Despite the opacity of the reasoning, the basic argument—that libraries face fiscal threats and threats to productivity—is clear.

Interestingly, however, the majority of articles in the 1976 proceedings do not focus on the theme of labor savings. Rather, attention is focused on the costs of various processes. These include costs of system design, computer supplies, and support; cost analysis of automating technical services; the economics of book catalog production and catalog conversion; cost analysis as a basis for decision-making; and the economics of automated circulation.

Only the passage of time could have made the 1976 proceedings possible. Experience with automated systems, especially in universities such as Ohio State and Cornell, as well as in cooperative enterprises such as OCLC, formed the empirical base for this level of managerial analysis. Without such experiences, talk of the economics of automation would have been only idle speculation.

But the newly automated library organization was not only concerned with internal costs. The outside world also presented its own fiscal threats and constraints. This was particularly true in the area of negotiating contracts with vendors. So it was that the 1977 proceedings devoted itself to negotiating for library automation.

The threat of poor negotiation was obvious: the price tag for automation was, and is, sufficiently high to threaten the fiscal viability of the institution and political viability of its director. In a way, the high price of automation had made library organizations acutely aware that they were, in fact, an open rather than a closed system, and that there were dramatic external factors that affected their production and survival. If management is the art of control, then attention would now have to be directed to controlling this uncertain environment. The "art of haggling" was given new meaning and importance, a new level of sophistication was required, and literature had become necessary to promote this sophistication.

J. L. Divilbiss (1977) noted that librarians still considered themselves at a disadvantage when negotiating for automation services for three reasons: one, the product and service were technically complex; two, the legal instruments were mysterious; and three, the vendor was a good deal more experienced in contract negotiation than the librarian (p. 1).

In this proceedings, the librarian learned the art of negotiating contracts with regional networks, automated circulation systems, and online data base services. The 1978 proceedings on failures in library automation offered an entirely new organizational dimension to the evolution of management concerns. Here, for the first time and in one place, managers could find out what mistakes others had made. This was, as issue editor F. W. Lancaster (1978) noted, "the other side of the coin." He reassured his readers that "it is perhaps not too surprising to find that the most abject failures are attributable more to management ineptness and bureaucratic bungling than to inadequacies in existing technology" (p. 1).

From a managerial perspective, it is hard to be reassured by this apologia for technology. But this harsh judgment was no doubt substantially accurate, and was a tacit recognition that Dobb's 1969 modified "Chinese proverb" was right: If the wrong people are in the right structure, the right structure will work in the wrong way.

But the litany of failure in the 1978 proceedings revealed not only the incompetence of managers. It also revealed the deficiencies of bureaucracies and the folly of believing unrealistic promises. Several articles, for example, discussed difficulties in dealing with governmental bureaucracies. One even listed twenty-eight steps that the state of California required the university to follow in order to purchase anything "even smelling like computers." Step 28 in this list of steps was: "Repeat steps 22 through 28" (Kountz, 1978, pp. 26-27).

James Corey (1978) from the University of Illinois at Urbana-Champaign discussed the organizational trials and tribulations of trying to develop an automated circulation system for the undergraduate library. He identified several traps into which the unsuspecting manager could fall. These included:

— not obtaining firm fiscal commitments from the administration,

— not obtaining enough money to get the system to the appropriate operational level,

— territorial conflicts among developers,

— taking too long a time for development,

— lack of understanding of the functions of the system, and

— developing a system that is too complex.

Corey's article highlights the need for reality orientation when buying into automated systems. The management of an automation project requires considerable attention to several factors requiring budgetary, political, and technical acumen. Most of general management can be a hit-or-miss process. Even when mistakes are made, there are often time and opportunity to correct them with minimal inconveniences. But it becomes painfully clear in reading the 1978 proceedings that automation projects take on a momentum very early in the development stage which, if improperly directed, can result in considerable and not easily reparable human and fiscal costs. There is a need for meticulous planning from the start by a realistic manager.

The realities of automation may be hard to grasp at first. Allan Veaner (1978) warns about the differences in actual and promised characteristics of automated systems development. He counsels the manager to beware of promised versus actual schedules, promised versus actual costs, and promised versus actual performance.

As a trilogy, the 1976-78 proceedings covered subjects of considerable contemporary interest: economics, contract negotiation, and reasons for failure. Discussion of these subjects formed a natural foundation for the evolving discussion of the relationship of automation to organizational behavior.

PHASE III. THE PROCEEDINGS OF 1983 AND 1985

The 1983 proceedings on competencies in library automation was a recognition that developing technologies had had a profound effect on library personnel. New technologies had stimulated the creation of new types of jobs and the evolution of old ones. The proceedings confirmed that sociological forces were at work, transforming the occupation of librarian to that of information professional. In the words of the clinic's editor, the 1983 proceedings considered "how professional roles and responsibilities have been and are being affected by technological change and what competencies are important in filling these roles" (Smith, 1983, p. 1).

Certainly, these were appropriate times to be asking such questions. In addition to the direct changes occurring in the information professions, more general developments in labor relations had created a definite need for clear definitions of the required knowledge, skill, and ability that should be possessed by workers. Equal employment opportunity court decisions were unambiguous in their assertion that individuals were to be judged on how their specific talents matched those required for a specific job. It should be noted, however, that the 1983 proceedings was not directed toward the practical managerial areas of job analysis or personnel selection; rather, it was a broader sociological study of how technology had changed the occupation of librarian.

The key questions for the 1983 proceedings were identified in an article by Jose-Marie Griffiths from King Research, Inc. She queried: "What are the current major trends affecting the library and information environment? . . . What do information professionals do? . . . What competencies are currently needed by information professionals to perform their functions and activities? What new competencies will be needed?" (Griffiths, 1983, p. 6).

Griffiths noted that information professionals now served in a wide variety of organizations, from libraries, information centers, and clearinghouses to database producers and distributors; from special collections and archives to information analysis and records centers. Providing a competent work force for these various agencies represented a substantial challenge. She argued that there must be cooperative planning between at least four groups: the information service organizations, education and training agencies, members of the research community, and professional societies.

Underlying this concern for competencies was an uncomfortable question: Do old librarians need new skills, or are completely new workers needed—a "new breed" of information professional? The notion is disquieting and strikes at the heart of not only the librarian but also the library school. Do library schools need new courses, or are completely new and different library schools needed?

Kathryn Luther Henderson (1983), writing on new competencies for technical services, found that what was needed for the future were librarians who were:

> . . . thinkers (with analytical minds), problem solvers, decision makers, and leaders . . . [they] must be inquisitive, curious, imaginative, and creative—they must be capable of managing, organizing, supervising, and communicating. And, at this particular time, the message that

> comes through is that they should be adaptable and flexible persons amenable to change . . . (p. 36)

By any standard, this is a librarian *par excellence.* Is this a new breed, or a refurbished version of the traditional librarian?

Danuta Nitecki (1983) from the University of Maryland Libraries, writing on new competencies for the public service librarian, noted that the librarian was being transformed into the "information consultant" or "information specialist" (p. 55). Richard Sweeney (1983), then Executive Director of the Columbus and Franklin County Public Library, seconded this chorus of rebirth and transformation. He exhorted that "librarians should not and must not be defined by a place—i.e., a library or even by a type of media such as the book" (p. 59). And Evelyn Daniel (1983) applied the *coup de grace* to the traditional librarian by reminding all that the profession is not one of librarianship but of "information professional" (p. 97).

It is a stimulating proceedings imbued with the assumption that new technology is here to stay, that it constitutes the defining environment for job analysis, and that people must adapt to technology's beneficence. Attention had shifted from the management of technology to the management of people. Only one article in the 1983 proceedings acknowledges the disruptive power that humans have over machines. Carolyn Gray (1983) from Brandeis University Library warns that optimism with automation must not ignore this power. The root of the problem, according to Gray, is that workers feel politically subordinated to the automated system. Failure to involve staff in the planning and decision-making process disenfranchises and can lead to sabotage—to what Gray refers to as "a new generation of Luddites" (p. 71).

Sufficient time had passed in the management of automation to know that people can be a major impediment to technological innovation. It is not surprising, then, that only two years later, the proceedings was entitled *Human Aspects of Automation: Helping Staff and Patrons Cope* (Shaw, 1985). This edition was in large part devoted to the management of people. It reflected the notion that organizational behavior had become a primary focus of automation management, and as such it was an index of the maturation of the discipline and its literature.

In this volume, major personnel issues that face managers of automation were examined. Topics such as "Resistance to Change," "Ergonomics," "Staffing," and "Planning and Implementation" were reviewed.

Why is it that people resist the machine? Sara Fine (1985), a psychologist whose research focused on human resistance to automation, reported that resistance is "alive and well among approximately 20 percent of our staff." (p. 4). Echoing Carolyn Gray's concern, Fine noted that human alienation results from attending to the computer, and this alienation is amplified when the decisions regarding automation are perceived as beyond human control. The result of such alienation is frustration and anger, which in turn produces apathy, sabotage, and employee turnover.

Of special significance in Fine's remarks, however, is the observation that resistance to automation can be a positive factor. It is probably not difficult for managers to remember a time when they wanted to institute a change, and there was at least one nay-sayer who kept pointing out one problem after another. It seemed a type of guerilla warfare, an attempt to wear managers' convictions down.

But the nay-sayer had raised some good points, which taken seriously and listened to might uncover potential problems which could be resolved before instituting the change. Resistors, according to Fine, provide a safeguard to the institution. The resistor must be respected and talked to rather than dismissed and ostracized.

It is an observation much appreciated by employees whose criticisms are often reduced by managers to the charge of provinciality or narrow-mindedness. But, although this is good advice for the manager, it has a ring of simplicity and näivete. It is like many management texts that talk about staff communication but say nothing about staff who simply will not listen. Neither Fine nor anyone else has an answer to that problem. Despite its flaws, Fine's perspective is refreshing and promotes respect for dissent.

Marvin Dainoff (1985) from the Center for Ergonomic Research explored another vexing area for the aspiring automator in the 1985 proceedings: ergonomics. Dainoff defines ergonomics as "an applied science concerned with the fit between people and the things (tools, equipment, environments) that people use" (p. 17). More commonly, it is the study of how technology affects the physical and, to a lesser extent, the psychological well-being of the worker. The potential physical damage from automation appears to be endless: backaches, neckaches, headaches, eye strain, and damage to the muscles of the arm and hand from inflexible chairs, improperly adjusted keyboards, and glaring Video Display Terminals (VDTs). Added to these are potential electro-magnetic dangers from VDTs. It may prove safer to live next door to a nuclear power station than to input OCLC records in one's local library! What is clear is that, in the current litigious climate in which any form of physical damage to the worker is subject to a claim of employer negligence, the area of ergonomics has become a necessary business of the automation manager.

The focus of both Fine and Dainoff is on the individual worker.

Fine deals with the domain of the mind, Dainoff with the domain of the body. But another issue for the automation manager is knowing what organizational factors are affected by automation, and how these factors affect the behavior of library workers. A useful overview and summary of these issues is given by Margaret Myers (1985):

1. *Staffing patterns:* Myers noted that the traditional separation of technical and public services functions was blurring. Automation had brought together files that were once physically separate. This blurring had created fuzziness to what used to be clearly defined work roles and work places among public and technical service employees. Such fuzziness could have serious effects on worker behavior. Research has suggested that high levels of role ambiguity can produce reductions in job satisfaction. In contrast, if the blurriness is perceived by employees as an opportunity for variety and challenge in the workplace, it can increase satisfaction. The current state of knowledge on this subject, as Myers notes, is imperfect. This leads to the second area identified by Myers, job analysis.
2. *Job analysis:* The introduction of new technologies, as noted earlier, has changed the nature of many jobs. This, in turn, affects job classification, and wage and salary structure of the organization. The challenge for the automation manager is to assess the impact of these changes on job classification. It must be remembered that reclassification may be perceived by the manager as a fruitful exercise in organizational rationalization, but may be perceived by the employee as an activity inclined to produce stress and conflict. In unionized environments, the potential misperception could be explosive.
3. *Professional support staff dynamics:* Myers notes that automation has freed some professionals from technical and clerical routines, and that these have been transferred to support staff; similarly, as professionals perform more sophisticated technological feats in automated searching, support staff have been allocated the additional responsibility of answering basic reference questions. Further, support staff are acquiring technical expertise not necessarily possessed by the professional librarian.

   How do these changes affect both the formal and informal relationships of professional and support staff? How are the authority and responsibility roles being redefined? Given that even in traditional libraries, professional and support staff relations are easily strained, will these new changes increase the status differentials or decrease them? Will a new organizational equilibrium be established? There is simply not enough information as yet to support an intelligent generality.

4. *Training:* Failure to train properly is a common problem that can produce devastating results for the organization and for the employee. However, it appears that, in the area of automation, library managers have recognized the seriousness of the issue. Myers notes that a study of 300 automation projects revealed that 50 percent of the costs incurred were for training. She warns managers not to underestimate the time required for staff to become acclimatized to new technology. It is good advice. Employees who fear that they are unable to learn or perform new tasks can produce resistance to change.
5. *Performance evaluation:* An effective system of performance evaluation must be based on sound information concerning the employee's performance. Automation has affected performance evaluation of the employee in an interesting way: it can provide what appears to be "objective" information. For example, in technical services, the number of items catalogued or processed by a particular employee can be determined. Similarly, the work of an employee can be checked against national or local standards to determine quality as well as quantity of work performed. The dark side, however, is that such monitoring by the supervisor could be interpreted as surveillance rather than supervision, and this could have serious impact on stress and morale factors in the organization.

CONCLUSION

The twenty-five years of proceedings, taken as a whole, reflect the recognition that library organizations are complex. The content of these proceedings reflects a logical progression from concern with the machine to concern with the human. There still remain, however, some deeper philosophical issues that have yet to be explored from a manager's perspective. Most notable is the issue of effectiveness. Is library automation real progress? David King (1986) recently argues that much of library automation may be "halfway technology," too costly and too complex to adequately solve the true problems of the library users, and that it may substitute problems that are defined and therefore more easily solved by the technology itself. Perhaps such philosophical speculation constitutes a fourth phase in the evolution of concerns for library management.

For the library manager, the development of automated systems has been both an occasion for celebration and a cause for trepidation. Today, as in the past, automation is considered to be an important new factor in the workplace. As a result, managers are still experiencing growing pains among staff and in organizational structure. In terms of the evolution of automation, managers may soon advance to a new stage—a stage in which automation will become less important and less significant because it will have been around a while. If a library has had an automated circulation system and an on-line catalog for ten or fifteen years, perhaps when that system is improved or changed the effects will be much less dramatic. Personnel will have accommodated, ergonomic factors will be considered as a natural part of change, and managers will know what to do before they do it. In this regard, the proceedings of the data processing clinics have contributed and will continue to contribute to that body of knowledge that will make these transitions smoother and more effective.

REFERENCES

Chapin, R. E. (1967). Administrative and economic considerations for library automation. In D. E. Carroll (Ed.), *Proceedings of the 1967 clinic on library applications of data processing* (Papers presented at the 5th Annual Clinic, 30 April-3 May 1967) (pp. 55-69). Urbana-Champaign: University of Illinois, Graduate School of Library Science.

Corey, J. (1979). The ups, down and demise of a library circulation system. In F. W. Lancaster (Ed.), *Problems and failures in library automation* (Papers presented at the 15th Annual Clinic on Library Applications of Data Processing, 23-26 April 1978) (pp. 35-49). Urbana-Champaign: University of Illinois, Graduate School of Library Science.

Dainoff, M. J. (1986). Learning from office automation: Ergonomics and human impact. In D. Shaw (Ed.), *Human aspects of library automation: Helping staff and patrons cope* (Papers presented at the 22nd Annual Clinic on Library Applications of Data Processing, 14-16 April 1985) (pp. 16-29). Urbana-Champaign: University of Illinois, Graduate School of Library and Information Science.

Daniel, E. H. (1983). Education matters. In L. C. Smith (Ed.), *Professional competencies—Technology and the librarian* (Papers presented at the 20th Annual Clinic on Library Applications of Data Processing, 24-26 April 1983) (pp. 97-108). Urbana-Champaign: University of Illinois, Graduate School of Library and Information Science.

Divilbiss, J. L. (1978). Introduction. In J. L. Divilbiss (Ed.), *Negotiating for computer services* (Papers presented at the 14th Annual Clinic on Library Applications of Data Processing, 24-27 April 1977) (p. 1). Urbana-Champaign: University of Illinois, Graduate School of Library Science.

Dobb, T. C. (1970). The administration and organization of data processing for the library as viewed from the computing centre. In D. E. Carroll (Ed.), *Proceedings of the 1969 clinic on library applications of data processing* (Papers presented at the 7th Annual Clinic, 27-30 April 1969) (pp. 75-80). Urbana-Champaign: University of Illinois, Graduate School of Library Science.

Fine, S. F. (1986). Terminal paralysis, or showdown at the interface. In D. Shaw (Ed.), *Human aspects of library automation: Helping staff and patrons cope* (Papers presented at the 22nd Annual Clinic on Library Applications of Data Processing, 14-16 April 1985) (pp. 3-15). Urbana-Champaign: University of Illinois, Graduate School of Library and Information Science.

Gray, C. M. (1983). Technology and the academic library staff, or, the resurgence of the Luddites. In L. C. Smith (Ed.), *Professional competencies—Technology and the librarian* (Papers presented at the 20th Annual Clinic on Library Applications of Data Processing, 24-26 April 1983) (pp. 69-76). Urbana-Champaign: University of Illinois, Graduate School of Library and Information Science.

Griffiths, J. M. (1983). Competency requirements for library and information science professionals. In L. C. Smith (Ed.), *Professional competencies—Technology and the librarian* (Papers presented at the 20th Annual Clinic on Library Applications of Data Processing, 24-26 April 1983) (pp. 4-11). Urbana-Champaign: University of Illinois, Graduate School of Library and Information Science.

Hage, E. B. (1967). An administrator's approach to automation at the Prince George's County (Maryland) Memorial Library. In D. E. Carroll (Ed.), *Proceedings of the 1967 clinic on library applications of data processing* (Papers presented at the 5th Annual Clinic, 30 April - 3 May 1967) (pp. 90-97). Urbana-Champaign: University of Illinois, Graduate School of Library Science.

Henderson, K. L. (1983). The new technology and competencies for "the most typical of activities of libraries": Technical services. In L. C. Smith (Ed.), *Professional competencies—Technology and the librarian* (Papers presented at the 20th Annual Clinic on Library Applications of Data Processing, 24-26 April 1983) (pp. 12-42). Urbana-Champaign: University of Illinois, Graduate School of Library and Information Science.

Kilgour, F. G. (1976). Economics of library computerization. In J. L. Divilbiss (Ed.), *The economics of library automation* (Papers presented at the 13th Annual Clinic on Library Applications of Data Processing, 25-28 April 1976) (pp. 3-9). Urbana-Champaign: University of Illinois, Graduate School of Library Science.

King, D. N. (1986). Library technology as "halfway" technology. *Proceedings of the ASIS Annual Meeting, 23*, 132-137.

Kountz, J. C. (1979). Problems of government bureaucracy when contracting for turnkey computer systems. In F. W. Lancaster (Ed.), *Problems and failures in library automation* (Papers presented at the 15th Annual Clinic on Library Applications of Data Processing, 23-26 April 1978) (pp. 23-24). Urbana-Champaign: University of Illinois, Graduate School of Library Science.

Lancaster, F. W. (1979). Introduction. In F. W. Lancaster (Ed.), *Problems and failures in library automation* (Papers presented at the 15th Annual Clinic on Library Applications of Data Processing, 23-26 April 1978) (p. 1). Urbana-Champaign: University of Illinois, Graduate School of Library Science.

Myers, M. (1986). Personnel considerations in library automation. In D. Shaw (Ed.), *Human aspects of library automation: Helping staff and patrons cope* (Papers presented at the 22nd Annual Clinic on Library Applications of Data Processing, 14-16 April 1985) (pp. 30-45). Urbana-Champaign: University of Illinois, Graduate School of Library and Information Science.

Nitecki, D. A. (1983). Competencies required of public services librarians to use new technologies. In L. C. Smith (Ed.), *Professional competencies—Technology and the librarian* (Papers presented at the 20th Annual Clinic on Library Applications of Data Processing, 24-26 April 1983) (pp. 43-57). Urbana-Champaign: University of Illinois, Graduate School of Library and Information Science.

Olsgaard, J. N. (1985). Automation as a socio-organizational agent of change: An evaluative literature review. *Information Technology and Libraries, 4*(1), 19-28.

Shaw, D. (Ed.). (1986). *Human aspects of library automation: Helping staff and patrons cope* (Papers presented at the 22nd Annual Clinic on Library Applications of Data Processing, 14-16 April 1985). Urbana-Champaign: University of Illinois, Graduate School of Library and Information Science.

Smith, L. C. (1983). Introduction. In L. C. Smith (Ed.), *Professional competencies—Technology and the librarian* (Papers presented at the 20th Annual Clinic on Library Applications of Data Processing, 24-26 April 1983) (pp. 1-3). Urbana-Champaign: University of Illinois, Graduate School of Library and Information Science.

Sweeney, R. T. (1983). The public librarian of the last years of the twentieth century.

In L. C. Smith (Ed.), *Professional competencies—Technology and the librarian* (Papers presented at the 20th Annual Clinic on Library Applications of Data Processing, 24-26 April 1983) (pp. 58-68). Urbana-Champaign: University of Illinois, Graduate School of Library and Information Science.

Veaner, A. B. (1979). What hath technology wrought? In F. W. Lancaster (Ed.), *Problems and failures in library automation* (Papers presented at the 15th Annual Clinic on Library Applications of Data Processing, 25-26 April 1978) (pp. 3-15). Urbana-Champaign: University of Illinois, Graduate School of Library Science.

W. D. PENNIMAN

Libraries and Information Systems Director
AT&T Bell Labs
Murray Hill, New Jersey

Systems Interfaces Revisited

INTRODUCTION

Lord Kelvin once said, "When you can measure what you are speaking about, and express it in numbers, you know something about it; but when you cannot express it in numbers, your knowledge is of a meagre and unsatisfactory kind . . ." (Thompson, 1889, p. 73). This article reviews four measures that relate significantly to current concern with information systems and their users. Two of these four measures have been discussed by Hal Becker (1986) in a delightful article concerning the world's growing capacity to store and transmit data.

In 490 B.C., the fastest way to send a message was through a human messenger running as fast and as far as he could. Often, he dropped dead upon completing the task (or was killed if the content of the message was displeasing). The data rate for that "system" was well under one word per minute (probably closer to 1/100th of a word per minute depending upon message length). Despite experiments with semaphore towers, carrier pigeons, and horseback riders, no really universal breakthrough came until the invention of telegraphy in the 1840s. With this technology, transmission rates achieved a level of about fifty words per minute. At present, we have reached transfer rates of one billion words per minute, and by the mid-1990s, the figure will exceed 100 trillion words per minute.

Storage capacity has made similar startling advances. In 4000 B.C., characters were stored on clay tablets at about one per cubic inch. Papyrus scrolls in intervening years improved the situation somewhat, but not until 1450 A.D. and the advent of the printing press using movable type did mass storage in the form of books jump to 500 characters per cubic inch. With high density electronic, magnetic, or optical technology this figure has been pushed to astounding levels, and by 2000 A.D. the capacity to store 125 billion characters per cubic inch will be realized.

Two other measures exist that are important to systems designers. Each has presented a significant barrier to effective information handling. One has only recently been resolved; the other remains a challenge today. Computation rates have been measured in instructions per second for some time. And for some time (since 5000 or 500 B.C.—authorities differ on its age), a single device held the world's speed record for computation when in the hands of an expert. The device was the abacus, and the rate was literally a handful of computations or instructions per second (probably two to four). It was not until the development of the electronic computer in the mid-1940s that this figure increased significantly. Then the rate of growth became phenomenal. In a few decades, the figure rose to one million, then ten-fifty million, and well beyond 100 million instructions per second (perhaps exceeding a trillion) is not unexpected with new computer architectures.

The final measure is that of human symbol processing capability. Around 4000-3500 B. C., when the first written language emerged, humans were capable of processing about 300 words or symbols per minute. At present, we are still capable of processing about 300 words per minute. Even with speed reading and listening devices and techniques, there are no order of magnitude changes in this figure as have been seen in the previous measures for transmission, storage and processing. The limit here is symbolic of an even greater barrier: our limited ability to make sense out of all the information being stored, processed, and transmitted. This last barrier is too often ignored in the design of information systems. Yet it is the most significant, for while costs of all the other components are dropping and capacities are increasing, this most vital component (the user) has a completely different set of characteristics. In general, if generalizations can be made concerning information systems users, they are becoming more expensive and less sophisticated. Therefore, any measure of system performance that does not include the user is ignoring the most crucial component.

Systems analysts like to draw boundaries around their world and measure within those boundaries, forgetting that the *user* must be inside those boundaries. But storage and retrieval are the *least* crucial of an information system's functions—particularly if what is being stored and retrieved is a brief surrogate of the information ultimately required.

BASIC SYSTEM REQUIREMENTS

Harry Goodwin (1959) of Battelle Memorial Institute published a checklist for the "ideal information service." The list is as follows (slightly paraphrased):

— to get information desired,

— at time it is desired (not before or after),

— in briefest form,

— in order of importance,

— with auxiliary information,

— and indications of reliability,

— and authority of information (source),

— to exert minimum effort,

— to be screened from undesired or untimely information, and

— to know negative results are reliable.

Although theoretically sound, experience has shown that such systems are not only difficult to implement but also extremely expensive.

When the U. S. Government decided to halt funding for information analysis centers in the 1970s and invest heavily in online systems instead, that decision put a burden on the computer that it was not yet ready to face. Much of the human intervention that would deliver the "ideal" system requirements was lost. The decision was clearly driven by economics. Labor costs were rising and the return-on-investment of the labor-intensive approach was not explicitly defined. This anecdote is not an argument to turn away from online systems, but to learn from the past.

Goodwin's article dismissed retrieval as a relatively small issue when compared to collection and dissemination of information. These are still crucial areas. Identifying potentially relevant information via an online search of a bibliographic database is important, but obtaining the truly relevant information contained in the source document is critical. That was recognized in 1959 and is recognized today (Penniman,

1987). Yet, full text delivery electronically for most documents is still not a reality.

Not only do computerized information systems ignore the most important system component, but they also address only a small piece of the overall problem.

Sherry Turkle (1984), author of a fascinating book on the next generation of users, states:

> When different people sit down at computers, even when they sit down at the same computer to do the "same" job, their styles of interacting with the machine are very different. . . . Some create worlds that are highly predictable and use their experiences in them to develop a sense of themselves as capable of exerting firm control. Others have different needs, different desires, and create worlds whose complexity is always on the verge of getting out of hand, worlds where they can feel themselves to be wizards of brinkmanship. (p. 15)

According to Webster, the term *individual* is applied to one "characterized by independence of thought and action." Since it is reasonable to assume that users of current information systems are, indeed, individuals in a literal sense, then a single model (mental or otherwise) of the user is unlikely to be adequate for design purposes.

Some shibboleths regarding system design and mental models include:

— The user's conceptual model is an integral part of the interface (McCann, 1983). (True, but so is the system designer's.)

— The user interface should be built around a conceptual model (McCann, 1983). (Yes, but whose should it be?)

— Training and documentation should be keyed to development of a conceptual model *by* the user (McCann, 1983). (And also *of* the user!)

— Visual language is quite efficient and universal but still shows some ambiguity (Ichikawa, 1983). (And it is the ambiguity that is the problem!)

The doubts raised are the author's and not those of the researchers cited above, but others have raised similar doubts regarding a single model for the user. One study indicated that different groups of users should be trained with different models (Borgman, 1982). Another indicated that explanation (in the context of building a model of the human explanation process) is a social process sensitive to context and individual differences of the people involved. These "individuals" bring particular assumptions and predispositions to the event which significantly influence what actually occurs (Goguen, Weiner & Linde, 1983).

A very old piece of research (McGuire & Stanley, 1971) offers an explanation for the vast differences that occur when individuals approach an interactive terminal and try to understand the underlying concept of the system while interacting with it. The findings are as follows:

— When confronted with a problem, individuals draw on past models. This is called model development.

— New models are built from old models as well as observation. These are called consolidation models.

— Similar experiences among different individuals elicit different models due to the consolidation process.

— After "understanding" (i.e., model formation) is achieved, disconfirmation is very difficult.

— Well-developed models are not systematically checked once formed by the user.

— Assumptions are made early and last long.

If McGuire and Stanley are correct, then system designers should be extremely cautious regarding assumptions about the universality of their command structures and user-friendly interfaces. Recent work indicates that the universality of search vocabulary or keywords (even for the most common of objects) is suspect as well (Furnas et al., 1987).

Today's Systems, Yesterday's Technology

Today's systems do not indicate that designers have learned from the past. The last two decades have seen the failure of major systems to reach their full (or even partial) potential. Videotext and interactive cable television are two painful examples where millions of dollars were invested and lost. CD-ROM is another potentially significant technology for information delivery. But is it a generic solution or a niche solution? Will it truly be a new papyrus (which, by the way, took a very long time to be used extensively), or will it confound information seekers by its lack of standard search techniques, user interfaces, and equipment? It is not at all uncommon to see multiple CD-ROM systems in a library today—each with a different user interface reflecting that database provider's view of "user friendly."

One wonders what a system would be like that was designed not to be "friendly" but rather "informative"—a system that provides direction clearly and concisely, that may not be "forgiving" but lets one know what it expects and offers correction when the right information is not provided. The analogy to this approach is an Army drill sergeant who is not friendly, but who teaches recruits very quickly what he wants.

The problems with CD-ROM (and its variations) and the multiplicity of user presentations or interfaces now available have already been referred to. While some internal standards exist, the user is still confronted with multiple systems, equipment, and search approaches. It would seem that this problem is perpetual.

There are major barriers to approval of the common command language standard by the National Information Standards Organization (NISO). In 1979, the author urged information professionals to promote an interim set of guidelines for user interface design based on available literature and pending the development of better guidelines as knowledge increases (Penniman, 1979). This was based on the assumption that interim guidelines are better than no guidelines at all. One effort to develop such guidelines is a 500-page MITRE report (Smith & Mosier, 1986) produced for the Air Force.

The AT&T Bell Laboratories has chosen two alternative approaches for direct access to their Library Network databases and services. The first is interactive and provides both a novice and skilled user interface. That service has grown from a few hundred users to over 5,000 in a little over a year (Penniman & Hawkins, 1988). Electronic mail is certainly not pioneering technology. In the AT&T Library Network, however, it has acquired a new meaning in the database and broader library service arena. Individuals throughout AT&T can access internal databases without ever learning a search command language or acquiring a logon to a remote machine. By executing the "library" command on their own machine, they are led through a series of steps that creates electronic mail to one of the AT&T databases (Waldstein, 1986). The results are returned to the user via electronic mail and no human interaction occurs. This "batch-oriented" system serves a much larger audience than can be supported simultaneously on the AT&T machine, and it provides a wider range of services as well. The user can submit document orders, reference questions, book or journal purchase requests and photocopy orders and subscribe to current awareness services—all via the modified electronic mail interface. The "library" command resides on most UNIX system machines throughout AT&T and communicates customer requests to the appropriate Library Network node where, in many cases, the message is processed without human intervention.

The Role of Behavioral Science

The user is the most expensive and least understood component of present day information systems, as already stated. Mental models can be useful (or misleading) in the design of information systems, as previously stated as well. All of this leads up to a framework for systems design that incorporates the following concepts:

— Individuals should be viewed as just that, and their differences taken into account in the design of systems.

— Subgroups of the user population should be studied for their similarities and their differences.

— The similarities across the total user population should be investigated.

— Past approaches have ignored individuals and subpopulations by continuing to focus attention on users as a homogenous population.

Most retrieval software design has assumed one or at most two (näive and experienced) user groups. Early software designs assumed large user populations with no need for skilled search intermediaries. Instead, a new profession arose because designers failed to recognize the skills and experience necessary to use their systems in an efficient manner.

The author has proposed a deceptively simple framework elsewhere (Penniman, 1985) built on the assumption that users can be studied just as other components of a system can and that performance measures can be developed from empirical studies of what users do (as a representation of what they think and what mental models are in play).

In various studies of user interaction conducted by the author, including one for the National Library of Medicine (Penniman, 1981), significant differences in the patterns of usage between groups of users have been measured. Even for homogeneous groups, a wide diversity of usage patterns emerged as more detail was added to the analysis. In other words, there was a high degree of individuality in the interactions observed, but such individuality was made up of short sequences of relatively common activities within a subgroup. These short sequences of activity provided insight into both the similarities and differences of user population subgroups and illustrated empirically how the users were actually performing in a measurable sense. The framework further leads to the following conclusions:

— Unique behavior and individual differences define the outer limits of the system and the degree to which it must adapt to a variety of user characteristics.

— Groupings of behaviors define shared models and will lead to the design of tailored components and command structures.

— Group behavior differences define the varieties of models that will be necessary to adequately serve all users.

And, above all, the framework illustrates that one must not assume a "universal" model of the user or system performance in its broadest sense will certainly be less than optimum.

A Broader Perspective

Although the previous framework appears to be a broad perspective of system design drawn from behavioral science, it is not a broad enough view. In looking back at a paper on networked databases that the author presented a decade ago, he was surprised to find that he was already beginning to suspect the larger complexity of the issues confronted regularly at present (Penniman, 1979). The author argued in favor of system boundaries that recognized the viewpoint of the user—not the systems designer. He also argued that the "system" boundary should encompass not only a search system and a document system, but also an education system (for users, intermediaries, and designers), and the bureaucratic and economic systems in which they reside. At the very least, system providers must understand the total system, respond to fundamental user requirements, use appropriate technology (not necessarily the most advanced—particularly when dealing with näive users), and establish links with other system components such as document production systems and document storage systems. Now more than ever, establishing links with economic and bureaucratic systems is essential.

Other historical data is instructive. A study was conducted for the National Science Foundation in which over 100 information innovations that failed were evaluated (Sweezy & Hopper, 1975). The results were startling: technology (or the lack thereof) accounted for just over one-quarter of the failures. The rest were due to management, marketing, capitalization, and organizational issues. In other words, the "technical system" isn't even half the battle.

CONCLUSION

It is appropriate to finish this article with a modern quote taken from a book called *The Ideology of the Information Age* (Qvortrup, 1987):

> People always dream about a better future, and our social system encourages this imaginative dreaming. The information society is one such social dream. Hence, technological creativity, as it expresses itself in information technology, is just as much social creativity. When discussing a possible better future, we must argue in social, not primarily technological terms. To make that future a reality, we have to act in social, not technological terms. (p. 134)

In other words, one can never forget the broader environment in which one operates. System boundaries must be realistic enough to make design practical and broad enough to make design realistic. That is the challenge facing systems designers today.

REFERENCES

Becker, H. (1986). Can users really absorb data at today's rates? *Data Communications, 15*(8), 117-193.

Borgman, C. L. (1982). Mental models: Ways of looking at a system. *Bulletin of the American Society for Information Science, 9*(2), 38-39.

Furnas, G. W.; Landauer, T. K.; Gomez, L. M.; & Dumais, S. T. (1987). The vocabulary problem in human-system communication. *Communications of the ACM, 30*(11), 964-971.

Goguen, J. A.; Weiner, J. L.; & Linde, C. (1983). Reasoning and national explanation. *International Journal of Man Machine Studies, 19*, 521-559.

Goodwin, H. B. (1959). Some thoughts on improved technical services. *Special Libraries, 50*(9), 443-446.

Ichikawa, T. (1983). Visualization of user interfaces introduction. *Proceedings Compsac 83* (The IEEE Computer Society's Seventh International Computer Software and Applications Conference, 7-11 November 1983) (pp. 338-9). Silver Springs, MD: IEEE Computer Society Press.

McCann, P. H. (1983). Development of the user-computer interface. *Computer Education, 7*(4), 189-196.

McGuire, M. T., & Stanley, J. (1971). Dyadic communication, verbal behavior, thinking, and understanding—II. Four studies. *Journal of Nervous and Mental Disease, 152*(4), 242-259.

Penniman, W. D. (1979). Designing the interface for the user. *Bulletin of the American Society for Information Science, 6*(2), 33.

Penniman, W. D. (1985). Information systems performance measurement—revisited. *Agenda for computing research: The challenge for creativity* (Proceedings of the 1985 ACM Computer Science Conference, 12-14 March 1983) (pp. 29-32). New York: ACM.

Penniman, W. D., & Hawkins, D. T. (1988). The Library Network at AT&T. *Science and Technology Libraries, 8*(2), 3-24.

Penniman, W. D. (1981). *Modeling and evaluation of on-line user behavior.* Final report to the National Library of Medicine under Extramural Program Grant No. NLM/EMP (R01LM0344401). Bethesda, MD: National Library of Medicine.
Penniman, W. D. (1978). Networked databases—A view from the middle. *Proceedings of the Thirteenth Meeting of the Geoscience Information Society, 9,* 39-54.
Penniman, W. D. (1987). Selling sizzle without the steak. *The EI Insider, 6*(Summer), 8.
Qvortrup, L. (1987). The information age: Ideal and reality. In J. D. Slack & F. Fejes (Eds.), *The ideology of the information age* (pp. 133-45). Norwood, N.J.: Ablex.
Smith, S. L., & Mosier, J. N. (1986). *Guidelines for designing user interface software.* MITRE Technical Report, MTR-10090 (AD-A 177 198). Bedford, MA: MITRE.
Sweezy, E. E., & Hopper, J. H. (1975). *Obstacles to innovation in the scientific and technical information services industry.* Final report to the National Science Foundation (PA Monograph 76-2). Washington, DC: Institute for Public Administration.
Thompson, W. (Lord Kelvin) (1889). *Popular lectures and addresses.* New York: Macmillan.
Turkle, S. (1984). *The second self: Computers and the human spirit.* New York: Simon and Schuster.
Waldstein, R. K. (1986). LIBRARY—An electronic ordering system. *Information Processing and Management, 22*(1), 39-44.

RALPH E. JOHNSON

Assistant Professor
Department of Computer Science
University of Illinois at Urbana-Champaign

A Consumer's Guide to User Interface Design

ABSTRACT

A computer system's user interface has a big impact on its acceptability and usefulness. There has been a lot of attention given to making software more user-friendly, but this has often generated more heat than light. This paper outlines the principles of good user interface design and discusses how to tell whether they are being followed or not.

INTRODUCTION

Software companies seem to have discovered the importance of good user interfaces. Every software package is promoted as being "user friendly," but many programs are still hard to use and easily confuse the user. A good user interface is more than windows, menus, or mice. Designing one takes careful attention to the needs of the user and to the ability of people to learn new ideas and predict the behavior of complex systems.

Software engineering education has traditionally placed little emphasis on user interface design, so many software designers ignore the user interface. Even when designers realize the importance of a good user interface, they may not be able to afford to design it properly. Moreover, there is no such thing as the perfect user interface, since an interface must be designed for a particular class of users. Any user interface will be ill-suited for some user. Thus, it is important for the buyer of a software package to be able to evaluate its user interface.

The commercial success of machines like the Macintosh has convinced software designers of the importance of a good user interface, and several books have been written recently that teach software engineers the principles of user interface design. Some of them are *The Elements of Friendly Software Design* (Heckel, 1989), a sermon on user interface design, *Designing the User Interface* (Schneiderman, 1986), which teaches human factors to software engineers, and *Designing User Interfaces for Software* (Dumas, 1988), which contains an extensive list of guidelines for software engineers to follow. This paper is aimed at the consumer of software, not the designer, and especially at librarians.

PRINCIPLES OF GOOD USER INTERFACE DESIGN

The most important word in "computer user interface" is user. Unless the designers of a computer system understand the needs and capabilities of the users, the system will not be easy to use. Whether a program features pull-down menus or function keys is not nearly as important as whether it was designed to be easy to learn by novices or to be efficient to operate by experts. Thus, a program's user interface must always be evaluated from the point of view of a particular group of users. However, as long as all of a computer system's users are human, they will have many characteristics in common.

The principles of good user interface design can be divided into three categories: (1) have a good idea; (2) put the user in control and (3) take the user's point of view. Unless users can understand the purpose and structure of an application, trying to make a good user interface for it is like trying to whitewash a garbage dump. Putting the user in control means making the application understandable and predictable, so that people can use it confidently and accurately. Taking the user's point of view means adapting the user's vocabulary, making the same assumptions that the user makes, and understanding the needs and characteristics of the user. The first and second sets of principles suggest ways of designing a user interface that make it better for all users, while the third relies on knowing details about the particular set of users.

Have a Good Idea

User interface design is a form of communication. Its purpose is to communicate the functions of the application program to the user. A pointless story well told is not satisfying, and a program that is poorly designed cannot be salvaged by a fancy interface. Even if the program addresses a real need, the solutions it offers might be too complex or solve the wrong part of the problem.

Keep It Simple

It is important for computer systems to be as simple as possible. Simple systems are easier to learn and understand, and users will make less mistakes and be more productive once they understand an application. While it is important for an application to have all the features needed to solve a complete set of problems, software engineers more often err on the side of providing too many features. The result is unreliable and unmanageable systems.

Software engineers always have a model of a program that consists of the kinds of data that the program uses and the ways in which that data can be manipulated. A simple design is one that minimizes the kinds of data and the kinds of operations on that data. A design is often made simpler by making the same operations apply to many different kinds of data. For example, there can be a single "print" operation instead of "print invoice," "print budget," and "print circulation report" commands. A design can also be made simpler by breaking complex operations into simpler ones. It is fine to let the system perform many steps at a time, but the user should be able to think in terms of the simple steps in order to predict what the program will do.

It is easy to be impressed by a complicated program with many features. However, simple designs often take a lot of work. Many complex systems have no overall design but are the result of adding feature after feature to an increasingly unstructured system. One should not be led astray by lists of features, but insist upon well thought out designs that integrate solutions to all the relevant problems.

Although a system may seem simple to the user, it need not have a simple internal organization. For example, users can add or delete text with ease using a word processor. However, the algorithms to format text might be complex. The user does not care about the complexity of the algorithms, only the results of adding or deleting text. As long as these results are easy to predict, the system seems simple.

Copy Reality

Good designs are often modeled after physical processes that users already understand. Most data processing systems copy the paper system that they replace, so their designs borrow the years of experience that went into the designs of the paper systems. A good example is the card catalog. Most library patrons understand (or think they understand) how the card catalog works, so it is not surprising that most computerized systems for assisting patrons are modeled after the card catalog. Doing so builds upon the users' existing knowledge.

Many of the problems for which applications are now being built have no such system to build upon, so it is essential that their designers come up with good metaphors to follow. A good example are the hypertext-based scholar's workstations (Meyrowitz, 1986). These systems integrate text and graphics and allow the user to browse through documents in many ways, adding comments and references to other documents. Designers try to manage the problems arising from letting many people change a single document by borrowing ideas from newspaper or magazine editing, but none of the solutions seem quite right. The major problem is that nobody has ever done anything like this before, so there are no solutions to reuse.

Given a good idea, a simple design is achieved by consistency of purpose, eliminating everything that goes against the central theme. This contributes to the illusion presented by the computer. Obvious mistakes in implementation destroy the illusion, so software bugs prevent a user interface from communicating the purpose of the application.

Put the User in Control

A user who is confused or who does not know what a program will do next is not in control. Users need an accurate mental model of a program to be able to predict what it will do. Programs whose form and function follow a simple model are easiest to learn and understand. Thus, it is very important that a program has a simple underlying model (the good idea) and that the user interface consistently expresses that model. Consistency is absolutely essential in making a program easy to understand, since it lets users predict the behavior of parts of the system that they have not seen from the parts that they have seen.

Providing Feedback

Users are most likely to learn the model if the application provides feedback through status messages and error messages. People learn best by having their mistakes corrected immediately. If a long period of time occurs between a user's command and the system's response, it may be difficult to figure out the reason for the response. Immediate feedback makes the user feel like he or she knows what is going on, so it enhances the feeling of control. When the computer pauses for a long time, the user is likely to wonder if the computer is broken. However, immediate feedback also improves the actual control that the user has over the application, since the user will be able to spot errors sooner and correct them faster.

Online Documentation

All users, even the most experienced, will sometimes need more information about how to use an application. Online help and documentation lets a user experiment with a system and learn about new parts of it. The user should be able to get help at any point in the application. This will eliminate the possibility of getting hopelessly confused and lost. Online documentation is not a replacement for printed documentation, since printed documentation is better as a tutorial and for learning about functions that have not been used before. However, a user should be able to complete most work without having to consult written documentation.

Online documentation is often not as well written as it should be. Since it is used more than the paper documentation, its quality should be at least as high.

Minimize Memorization

People are much better at recognizing detailed information than they are at recalling it. Users should be able to select commands and options from lists and so use their ability to recognize correct choices. This is the primary advantage of menus over commands to control an application.

Screen layouts should be designed to minimize memorization. If the user needs to combine two types of information to make a decision, it should be possible to display them both at the same time. One sign that this principle is violated is if users are forced to repeatedly flip between screens to compare results.

Prompts for data entry should describe any special formats that are required. For example, a prompt for a date should describe whether a purely numeric (mm/dd/yy) or an alphanumeric (dd MMM yy) entry is required. Computers are much less flexible than people at adapting to different formats, so it is only fair that the program remind the user of its particular requirements.

Online help and good status messages are very useful at jogging the user's memory. Each screen should be labeled so the users always know where they are.

Consistency

Consistency is extremely important in reducing the need to memorize. The same command should mean the same thing in every part of the application. Systems like UNIX that are a collection of different programs are often very inconsistent in their commands, since these programs have many different authors (Norman, 1981). Depending on the program, the command to leave a UNIX program might be *x* (exit), *q* (quit), bye, or ^*D* (the end of file marker). It is no wonder that new users find UNIX confusing.

A consistent vocabulary will reduce the number of new concepts that the user must learn. A computer neophyte is likely to get confused if told to *press* a key, then to *type* a key, then to *hit* a key. All vocabulary should be consistent, including that of printed documentation, online documentation, and messages.

In the same way, data formats for dollar amounts or dates should be the same throughout an application. The same formats that are used in messages should be used in data entry. Complex data, like author/title/publisher for a book, should be formatted the same every time they are displayed.

Consistency is very hard to achieve, and will not be achieved unless it is one of the designers' main goals. User interface designers must have a detailed set of rules for what the user interface will look like, usually in the form of a user interface design handbook. This handbook will describe screen layouts, data entry procedures, command structures, online help, etc.

Sometimes a user interface design handbook is made public. For example, Apple has a set of rules for the Macintosh (Apple Inc., 1987) that describes what a standard Macintosh application looks like. The result of having a set of design guidelines is that a new Macintosh program can often be used without reading its manual.

It would be very helpful if each software package would come with a copy of the user interface design handbook used to build it. For one thing, those systems without a handbook would be obvious and could be avoided. Unfortunately, most companies consider their handbook to be proprietary. The user manual should be able to provide a general description of the design rules which one can use to learn how to use the application and to determine whether a consistent set of rules was used to design the user interface. It is important to find out whether or not an application was designed with a user interface design handbook, since it is nearly impossible for a large system that was built by many people to have a consistent user interface without one.

Hide Unnecessary Details

Lurking beneath the surface of every application program is a hoard of details that are completely uninteresting to the user, not least of which are the computer's hardware and operating system. These details should be hidden from the user. The user should never see a message from the operating system complaining about bad file names or lack of disc space. Instead, the application should catch these errors itself or explain them in the user's own language. A user interface should communicate the purpose of the application, not the peculiarities of the system software that it is using.

Users do not care about how the application is implemented, either.

For example, the original catalog searching program at the University of Illinois at Urbana-Champaign forces users making title searches to type in the first four letters of the first significant word of the title and the first five letters of the second. This is presumably done because of the way that the database is organized, but users don't care about this detail. Users should be able to enter the entire title and let the computer figure out which words are important.

Take the User's Point of View

There are many ways to characterize groups of users. An information retrieval system might be designed for professionals who plan to use it for an hour or more every day. These users are willing to spend several hours learning to use the system, but need an interface that makes good use of their time once they have mastered it. On the other extreme, the system might be designed for members of the general public who will not use it more than a few times a year. These users need an interface that leads them through the functions of the program without bewildering them. University students might make up a third group, since they will usually be familiar with computers, though not with the principles of information retrieval, and can use that knowledge to learn about new systems.

Some computer systems with a reputation for poor user interfaces actually have interfaces that are well designed for a small, atypical group of users. A good example is UNIX, which is an operating system developed by a few computer scientists within Bell Laboratories and which is favored by computer science departments. According to legend, the large number of cryptic two- and three-letter commands were chosen because the early designers were slow typists and wanted to be able to quickly enter a large number of commands. Thus, the same features that make UNIX hard to use by the general public made it well suited for its original set of users.

Adopt the User's Vocabulary

The importance of speaking the user's language is most clear for software being shipped to another country. However, even among English speakers, different people have quite different vocabularies. In particular, software designers are usually quite different from users and have a different vocabulary. Thus, many programs use computer jargon to communicate with those who know little about computers. Just as bad is an accounting program that uses accountant's jargon to communicate with the new manager of a small business who is trying to learn to use it, or an online catalog that uses librarian's jargon to communicate with patrons. Inappropriate jargon is worse than meaningless, because it can make the user feel stupid and frustrated.

Jargon is not necessarily bad. If a group of users share a common language, then programs that use that language will be able to communicate with them better. However, users outside that group will be confused by the jargon. For example, an expert system that is helping a doctor diagnose a disease must use the medical terms with which the doctor is familiar, but these terms will be meaningless to most other people.

Adopt the User's Assumptions

It is easiest to communicate with someone with whom one shares a common set of knowledge and beliefs. The main advantage to basing computer systems on some preexisting paper system is that users will already know the kinds of information stored in the computer and what can be done with it. Thus, users will start off with a close approximation to the application model. This will decrease training time and anxiety level of new users.

The expertise of users can be measured along several dimensions, such as their knowledge of computers and their knowledge of the application domain. Patrons of a computer science library can be expected to be experts on operating a computer, but they will not know any more about how a library works than other library patrons.

Hidden assumptions often cause problems. For example, people have preconceived ideas about many colors. Auto drivers think of red as meaning stop or danger, chemical engineers think of red as meaning hot, and electrical engineers use red lights to mean that the electrical power is on. It is easy for users to place their own interpretation on the use of colors even when this goes against the intention of the user interface designer.

Understand the Needs and Characteristics of the User

People rarely want to use a program just because of its fancy interface. Instead, they have a job to do and think that the program can help them do it. For example, most students use the library to write papers for classes. They usually know the subject of the paper, but not the books on that subject. The traditional card catalog, which puts author and title at the same priority as the subject, is not suited to their needs.

Typing skills can have a big impact on how easy a particular program is to use. Some programs almost eliminate typing by using menus and touch screens or mice. A skilled typist will find a touch screen interface considerably slower than one based on a keyboard, but most other people will prefer to use a keyboard as little as possible.

Computer systems in libraries have many different kinds of users. Patrons of public libraries can be characterized by a general unfamiliarity with the intricacies of either libraries or computers, but there is little else they have in common. Some are poor typists, some are hard of hearing, some are color blind, some are quite young, and others are elderly. Patrons of a university library will be more homogeneous, but even in a university library the user interface for patrons is probably the biggest challenge for the user interface designer.

There are other important groups of users in a library. Professional librarians are likely to be willing to spend time to learn the details of a program, and might be interested in user interfaces that they can customize to make as efficient as possible. Data entry clerks are likely to be skilled typists but will not know much about computers. Some patrons, such as graduate students or others doing research, might be more concerned with the efficiency of the user interface and less concerned about its ease of use. It is hard for one interface to satisfy all these users.

TRENDS IN USER-INTERFACE DESIGN

The last few years have seen an explosion in the amount of software developed to be easy to use. This is due in large part to the popularity of personal computers, which have introduced many new people to computers, making it more important for software to be easy to use and providing a large market for good user interfaces. This has resulted in several new ideas in user interface design. Some, like direct manipulation and customizable interfaces, are certain to be even more important in the future.

Menus

A set of user interface guidelines published in 1975 did not even mention menus (Engel & Granda, 1975). Now they are almost synonymous with user-friendly software. Menus rely upon the fact that it is easier to recognize the name of a desired command than it is to remember it. Thus, menus are especially helpful for novice users. However, even experts will appreciate menus for those parts of a program that are rarely used.

Menus alone will not make a program easy to use. In fact, menus can be easily misused. Menus should not be too large or too small, they should be well formatted, menu items should have well-chosen names, they should be given some kind of logical organization with spaces between the different categories, and there should be standard ways to exit a menu without selecting any of the choices. In general, if one has used menus that are implemented well, then poorly implemented menus will be obvious. If one did not design the menu, then poorly chosen menu names and poorly organized menus will confuse the user, but the designer of such menus never seems to be bothered by them.

There are several kinds of menus. One kind is a list of options labeled with letters. The user types in the label of an option to select it. These kinds of menus can be used on nearly any kind of terminal. Other menus are sets of icons or lists of words that can be selected by a pointing device. These kinds of menus require special graphics terminals with pointing devices. Although these kinds of menus seem very different, they are all effective at reducing memorization.

The major problem with menus is that they can slow down experts. Experts prefer user interfaces that allow them to issue commands from the keyboard without having to navigate through a set of menus. Unless the terminals are fast, displaying menus can take a lot of time. Thus, menu systems often allow users to also enter commands from the keyboard or to type a string of item names corresponding to a series of selections from the menus. These interfaces can be easy to use by novices and still efficient for experts.

Multiple Interfaces

One obvious solution to the problem of having to cater to the needs of many kinds of users is to provide an interface for each kind. This was very hard to do in earlier applications because the user interface was implemented as an integral part of the application. A recent trend is to separate the construction of the user interface from that of the rest of the application. This is useful because user interfaces need to be revised more often than other parts of the application, because it makes it easier for the user interface to be designed by human factors experts, and because it makes it easier to provide several interfaces for a single application.

The Macintosh places all messages to the user in a special "resource" file. By providing a different resource file, all the messages in the application can be changed. This makes it easy to provide foreign language versions of applications or even to provide slightly different messages for different kinds of users.

Another way to provide multiple interfaces is to provide several different modes for different levels of user. For example, Microsoft Word has a "short menu" mode and a "long menu" mode. Short menus contain only the most used commands, keeping new users from becoming confused by too many choices. Long names provide the experienced user with the entire set of commands.

Customizable Interfaces

A customizable user interface is one that the user can change. Users might be able to change the name of menu items or function keys, to add new menu items or function keys that invoke some combination of existing commands, or to invent new commands. Different user interfaces vary widely in their customizability, but even the ability to change command names, menu names, or messages lets users with different vocabularies coexist.

Weissman discusses customizable user interfaces in depth in this proceedings. Customizable user interfaces are most valuable to expert users, since they can afford to spend the time to customize and often are more particular about it. Even novices can benefit from customizable interfaces, since an expert can easily modify the interface to be more suitable for a particular group of novice users.

Direct Manipulation

A direct manipulation interface provides the illusion that the user is physically manipulating the data that is inside the computer. The model of the application becomes so real to the users that there is little translation from the user's intentions to the actions needed to carry out the intentions. Direct manipulation interfaces often make heavy use of graphics and pointing devices, letting the user select and move objects that appear on the computer screen.

When a direct manipulation interface is combined with an application model that the user already understands, it provides a very easy to use interface. Electronic spreadsheets and word processing programs are both good examples. Paper spreadsheets have long been used for analyzing budgets and for accounting. Electronic spreadsheets model the paper spreadsheets fairly directly except that they automatically calculate the spreadsheet according to rules entered by the user. Changing the numbers or calculation rules in an electronic spreadsheet causes an immediate change in the numbers that depend on them. This immediate feedback is an important part of direct manipulation systems.

Early spreadsheet programs like VISICALC and Lotus 1-2-3 used function keys to select and modify individual cells, which meant that it took a little time for a new user to figure out how to use these programs. Users of newer programs like Excel can select cells by pointing at them with a mouse and can modify them by typing in new values. The graphical display makes the electronic version of the spreadsheet look exactly like the paper version, so new users have no problem figuring out what is going on. The improved interface makes manipulating the newer spreadsheets even more direct.

Word processing programs provide a direct manipulation interface when they display the document being created in a form as much like the eventual printed form as possible. Text is formatted as it is added, so the display always shows the current state of the document. Text is deleted or moved by selecting it with the pointing device. Many complicated operations, such as changing the size of characters or the fonts being used, are invoked by complicated techniques. However, the basic editing operations are always very simple and intuitive. Using these systems is like rearranging words in a magic book that not only formats the text automatically, but complains when words are misspelled.

A direct manipulation interface to the cataloging database could model the card catalogs with which patrons are familiar. Users could fill in part of a card and have the computer find all cards that match the information that was filled in. Users could move the images of interesting cards to one side of the screen, throw away the images of cards that were not interesting, copy information from interesting cards into blank cards to request more items from the catalog, and end up with sets of cards that they can print out or use to check out the corresponding documents.

New Hardware

It is clear that graphics is becoming more and more important to personal computer users. New personal computers all have graphics capabilities, and the cheapest of these machines are about the same price as some character-only display terminals.

While color is important to people and will eventually help make user interfaces more effective, there are only a few special areas in which it is known how to make good use of color graphics. A color interface is obviously useful in preparing color charts or illustrations. Integrated circuit design and some other computer-aided design systems make effective user of color. However, many attempts to use color result in garish, hard-to-read displays.

There are a few ways in which color is useful in text-based applications. Color works well to attract the user's attention, especially to changes in status. It allows the user to spot related items that are widely scattered on the screen. It also can help format a densely packed display. However, color displays are more expensive than black and white displays, and it is not at all obvious that color displays are cost effective in the library environment.

Pointing devices like mice or track-balls can be used in conjunction with graphics to reduce the use of the keyboard and make it possible for people to use the computer with no training at all. The need for memorization can be reduced to a minimum, and graphics can be used to provide excellent feedback. Thus, systems that combine simple graphics and pointing devices will become more common.

Touch screens have been available for a long time, but software technology has finally advanced to the point where they can be useful. Using a touch screen for a long time will make most people's arms tired, but the screens are easy and intuitive to use for novices. Thus, there will probably be a place in libraries for touch screens.

An interesting new idea that shows much potential is the data glove. This is a glove with sensors in it that the computer can use to determine the position of the user's hand. The computer can tell whether the hand in the glove is pointing, grabbing, or pushing items on the display. The result is a remarkably powerful interface. However, full development of this kind of interface is several years away.

Voice output is becoming more popular, showing up in videogames and personal computers. Voice input is far less capable. Some voice input systems can understand a small fixed vocabulary for a general population, or be trained for a large vocabulary for a single person, but none are even close to being suitable for use by library patrons.

FINDING A GOOD USER INTERFACE

It is easiest to buy or build software for oneself. Since the buyer and the user are the same person, if one buys what one enjoys using, then one will usually make the right choice. However, no new user is an expert at using the program, so if speed for an expert is important, then one should not buy a program based on how one liked using it for an hour in the store.

Unfortunately, software is often bought or designed by someone other than its final user. People usually buy or design software that they would like to use, even when the main users of the software will be other people. Since a user interface that is good for one person might be bad for another, the result is frequently software that is poorly suited for its users.

Software for the personal computer mass market is reviewed extensively by many magazines. These reviewers usually examine closely how easy the application is to use and to learn. Inconsistencies in the user interface, lack of documentation, or confusing commands will provoke negative comments from the reviewers. Moreover, since this kind of software is so widely used and has so much competition, it is relatively easy to find people who have used several competing systems and can provide an accurate description of their relative strengths and weaknesses. However, different users have different needs, and one may need customized software or software for less popular computers. Thus, it is important to know how to evaluate a user interface.

The amount of effort one puts into finding a system with a good user interface depends on the importance of making the right choice. An extensive evaluation of a program can be expensive and is only necessary if the software is expensive, will play a crucial role in one's organization, or will be used by lots of people. Often the cost of training and recovering from a mistaken purchase is much more than the cost of the purchase itself, so one should not underestimate the importance of making the right choice.

Testing

Testing a user interface follows the methodology of the social sciences more than that of engineering. Perhaps this is one of the reasons why so many software systems have poor user interfaces. Their designers often have an engineering background and have not been trained to analyze human behavior.

First, a representative sample of users must be gathered. Only a few users of each type are needed, but each kind of user must be represented. The exact number of people depends on the difficulty of performing the tests and the importance of making the right choice. Finding the right set of users is probably the hardest part of testing a user interface, and requires a careful analysis of the user population and how they will use the system.

In place must be a defined list of representative tasks for the users to perform. Groups of people that differ in the reasons they want to use the system or in the functions they will perform with it will need different lists of tasks.

Given a set of users and a list of tasks, the quality of a user interface can be measured by the amount of time it takes each user to successfully complete his or her list of tasks. Although there are other properties that could be measured besides the speed of using the system, it turns out that most of them are included in measuring speed. For example, if an interface causes users to make many errors, a lot of time will be spent making corrections, and it will take longer to complete a set of tasks. Speed is a poor measure if many users never complete their tasks as is discussed by Leslie Edmonds in this proceedings.

It can be hard to evaluate a user interface designed for experts. First, it takes time to train users to be experts so that the test is valid. Second, experts will probably use the computer for hours at a time, so ergonomic properties of the user interface are important for them and need to be evaluated. Third, systems used by experts are usually more complex, require more documentation, and can make use of features such as customization of the user interface that are not as useful for novices. Thus, expert users still have a much larger list of tasks than novices. It takes more time to evaluate a user interface for experts than one for novices.

This form of evaluation is not adequate for system designers, who need to diagnose problems with a user interface so that problems can be corrected. However, it works well for deciding which existing system is easiest to use.

Questions

There are a number of questions that help to evaluate the user interface of an application. Many of these questions cannot have any absolute answer, so it is best to use them to compare competing systems. The following list is probably too short, but it provides a good indication of the kinds of questions that are important.

— For whom is the system designed?
— Is the system too complex?
— Is the basic idea well-known to the intended users?
— Is there enough feedback?
— Is there high quality online documentation?
— Is the user forced to memorize commands? Is the user forced to flip from one screen to another?
— Is the system consistent?
— Is there a set of rules to ensure consistency?
— Are implementation details hidden from the user?
— Is the vocabulary of the program the same as that of the users?
— Are menus and displays well organized?
— Are command names easy to understand?

Answering these questions can help one understand why a particular system is hard to use and will be helpful in persuading the designers to fix their mistakes. However, the answers will be subjective and are no substitute for objective measurements. Good user interfaces are important because they make the users more productive, so measuring the productivity of users is the ultimate test of a user interface.

CONCLUSION

User interface design has improved significantly in the last few years. However, designing a good user interface is still difficult and expensive. Software vendors are not likely to design good user interfaces unless their customers demand it. If expectations of software quality are increased, then software vendors will be forced to provide user interfaces that are efficient and easy to use.

REFERENCES

Apple, Inc. (1987). *Human interface guidelines.* Reading, MA: Addison-Wesley.

Dumas, J. S. (1988). *Designing user interfaces for software.* Englewood Cliffs, NJ: Prentice Hall.

Engel, S., & Granda, R. (1975). *Guidelines for man/display interfaces.* IBM Technical Report TR00.2720. Yorktown Heights, NY: IBM Research Division, T. J. Watson Research Center.

Heckel, P. (1984). *The elements of friendly software design.* New York: Warner Books.

Meyrowitz, N. (1986). Intermedia: The architecture and construction of an object-oriented hypermedia system and application framework. *SIGPLAN Notices, 21*(11), 186-201.

Norman, D. A. (1981). The trouble with UNIX. *Datamation, 27*(12), 139-150.

Schneiderman, B. (1987). *Designing the user interface: Strategies for effective human-computer interaction.* Reading, MA: Addison-Wesley.

JESSICA R. WEISSMAN

Computer-Based Training Manager
James Martin Associates
Reston, Virginia

Have it Your Way: What Happens When Users Control the Interface

INTRODUCTION

When computers were new, nobody had fun with them except, possibly, the people who created them. The computers themselves were locked away in special rooms and not everybody had access to them. Users, even the most serious of programmers, spent many hours going over their programs and other input just to be sure it was perfect. The act of programming was carried out at desks, using paper and pencils.

When the programmer was finished, another group of people translated the program into a set of punched cards. This was a particularly slippery and risk-prone embodiment of the hours of work the programmer had already put in. When the card deck was ready, the programmer or someone else took the stack of cards to an input clerk.

The input clerk had tremendous power. She (they were mostly women) decided whose jobs could jump ahead in the line. Hours later, the programmer got back his output, generally in the form of a printout. If everything went well and there were no mistakes of form or logic, the results would be useful. If either the programmer or the keypuncher made even one tiny slip, all the hours of work and waiting went to waste. Even if the mistake was a trivial or easily discovered one, the programmer had to wait for his next turn to have his program run. In many installations, programmers got only two or three runs per day.

The whole system was geared to make the computer important and the user unimportant. In fact, users in the current sense did not exist. Essentially, everybody who used a computer was a programmer or a keypunch operator or some other kind of specialist. The computer's time was considered valuable, so valuable that it was measured in expensive seconds. The programmer could work for hours in order to save the computer a few seconds.

There was a great separation between the computer and the users, both physical and psychological. The programmer worked only at a distance with the material of the computer world. Except for a few visionaries like Vannevar Bush, nobody foresaw today's highly interactive computer world.

Nevertheless, the rewards of doing something new and creative, and the thrill of getting the computer to do something were there. People like Grace Hopper had plenty of fun, and so did the other computer creators. Within a few years, interactive computing began to flourish, and the creation of software that posited an active user was common. Still, the users had to do what the programmer wanted them to do, within the limits set by the programmer. Clearly, someone else was in control.

The subject of this paper is the way that users get to share what programmers have—the feeling of control over and comfort with their machines. In the dozen years of the personal computer's existence, the role of the user has moved from grateful but miserable wretch or computer whiz to kingpin. Several types of user-modifiable interfaces will be discussed, including: keyboard redefinition and macro programs; macro facilities built into spreadsheet and word-processing programs; "work" menus created by the user; full-blown customizable interfaces; and Apple's HyperCard program for the Macintosh. How each kind of user-controlled interface empowers the user and changes his or her relation to the computer will be discussed.

The existence of and the acceptance by users of all these interface control tools both enabled and marked a fundamental change in the role of the user. Users have become more like programmers while still remaining users. Without being highly technical, without thinking of themselves as "computer whizzes," people who use computers for productive daily work as well as pleasure have come to feel that they are in control of the computer. By acknowledging user need to control the interface, programmers have divested themselves of some of their specialness and shared some of their satisfactions with users.

Eventually, computers became more common, and more work took place at terminals, first printing terminals and, eventually, video terminals. The rise of video terminals drew the user closer to the computer in two ways. First, it made the programming process more interactive and immediate. No one mediated the programmer's contact with the computer. Second, it made interface important. Programmers had to use keys and screen displays to get anything to happen.

Video terminals made possible interactive programs intended for ordinary users rather than programmers. People who did not want to devote their lives to computers began to use computers for tasks ranging from data entry to financial modelling. A separate class of computer-like machines, dedicated word processors, came into being. These had no programmers, only users whose goal in using the machine had nothing to do with the computer itself. However, even dedicated word processors were run by key operators, who had to know far more about the operations of the machine than the rest of the users did.

What makes computers attractive to programmers? What do programmers like about programming? These questions are related to ordinary dedicated programmers who bring some passion to their work. The discussion is based on the author's own experience as a programmer, conversations with other programmers and users, and from such accounts of programmers' experience as Sherry Turkle's *The Second Self* (1984). A strong common thread is the experience of control. For programmers, the computer is a place they can control and understand. The computer does what they want it to do, the way they want it to. Of course, this is an ideal, and it takes plenty of work to get the computer to do the right thing. But the programmer is in control. He or she determines what the computer will do, and the computer always does what the programmer tells it to. The problem is that only the computer knows what it has actually been told to do. Hence, the frustration of debugging.

Closely related to control is the urge to personalize the computer. If the computer is the world the programmer interacts with, the programmer wants to make it his or her world. There are plenty of equivalents in the computer world of those paintings one sees on the sides of vans. Programmers can determine the wording of the prompt they see. They add their own twists to the operating system and the editor they use. Many of the small twists they add have about the same function as putting up pictures on the door of an anonymous dormitory room—they show who is the owner.

Programmers also share an urge to fiddle and play. Lots of programmers also have hobbies like ham radio and model railroads—the kinds of toys that permit and reward endless fiddling. If one doesn't want to work but still wants to be in contact with the computer, there are plenty of housekeeping things to do. The IBM-PC owner can alphabetize files in directories, and the Macintosh owner can move files around between folders, and it looks like work. For people who control a larger system, the scope for play is much larger. One of the best accounts of this was written by Ray Ozzie (1986), formerly a PLATO system programmer who developed the Symphony program for Lotus. His almost poetic description of the fun to be had with the PLATO system when the users were all home in bed can be found in a book called *Programmers At Work* (Lammers, 1986), a hymn to creativity and fun that makes inspiring reading.

Another thing that programmers like is the feeling of immediate gratification. Despite the problems involved in debugging, when the program works, it works. The rewards are small and constant. Frustration only increases the intensity of the reward when the program finally works.

The rise of the personal computer gave programmers the ideal field in which to create interactive programs. Interactive here means programs that are built on constant communication between the user and the machine. The first personal computers required the utmost in understanding from their users. One had to be not only a programmer but an electronics expert to get anything out of them. They came in kit form and had to be assembled and then programmed by flipping switches on the outside of the case. Even when BASIC came along and people could program in a traditional manner, there were no plain users. Everyone was a programmer. Ease of use was not a goal.

The first few commercial programs and games hardly altered the situation. Only when VisiCalc gave ordinary people a reason to use the computer did a large class of nonprogramming users arise. These pioneers experienced a lot of difficulty. Computers were still tough to use and computer programs still rigid and mysterious. People joined user groups in order to get enough information to use the machines properly, and to share their computer frustration.

When the personal computer became more commonly used for productivity, the class of pure users arose. These users did not see the computer as a challenge to their skill and understanding. They did not want to play with their computers. They wanted to use the computer as a kind of typewriter/math machine/toaster. So interfaces were made easier for them. Programmers and designers began to take the idea of ease of use seriously, seeing that the audience was no longer their fellow hobbyists.

After a few more years, the sophistication of the users rose, and they began to see the computer as something they could control. Enough experience with computers gave users sufficient understanding of the computer to imagine better ways of working with it. The availability of interface-altering tools brought the users' concept of what they did with a computer much closer to the programmer's idea. Without having to do programming-surgery on the programs, users could make programs behave the way they wanted them to. Users could master the computer world they lived in without having to become experts. A new kind of equality arose between users and programmers.

Several types of user-modifiable interface tools have arisen over time. Some of these are:

— keyboard redefinition and macro programs (such as Prokey, Keyworks, Superkey, Tempo, and Quickeys);

— macro facilities built into spreadsheet and word-processing programs;

— "work" menus where the user can promote any item buried in the regular menus to a special top-line menu;

— full-blown customizable interfaces as found in programs such as Borland's Quattro and Sprint, where the user can create his or her own set of menus using an interface-creation language not unlike a macro language; and

— Apple's Hypercard program for the Macintosh, a program that consists almost entirely of interface.

This list includes only methods where the user can control the interface within a program. Another large but amorphous class of user-controlled software customizes the computer system itself. Into this class fall the hundreds of utility programs that can alter the directories searched by the computer or allow the user to review what just scrolled by; or blank the computer screen after a period of inactivity; or make the irritating, blinking cursor into a friendly, steady, solid block; or supplement the Macintosh Finder with a more traditional method of selecting, moving, and deleting files. These are not discussed because they are less user-oriented. They allow the kinds of personalization that programmers enjoy, and some of them make life much easier for those who employ them. But they don't empower the user in the same way as the other tools being discussed here.

KEYBOARD REDEFINITION AND MACRO PROGRAMS

Keyboard redefinition and macro programs came to the personal computer in late 1983. Well-known programs of this type include Keyworks, Prokey, and Superkey for the IBM PC; and Tempo and Quickeys for the Macintosh. These programs use a single mechanism to do several things. They allow the user to decide what will happen when a given key is pressed. One can decide that when one presses, say, CTRL-Y, the computer will hear, instead, CTRL-Z. If a familiar program does something harmless like open up a dictionary when CTRL-Y is pressed, but a new program deletes the entire page on the same keypress, it is in the user's interest to block off the now disastrous CTRL-Y. So users made programs mimic each other, and protected themselves from destructive mistakes.

If accented letters or other special characters that require horrendous finger twists are regularly needed, the keyboard can be redefined so that some unused combination gives the needed letter. In other words, the user decides what is important enough to be accessible. Users can also assign whole strings of keypresses to a single keypress. A single keystroke could issue the commands to, say, type a standard letter closing and signature, open a new worksheet and fill in standard headings, or close one program and open another. In fact, keyboard macros can automate all kinds of procedures. Users can essentially add new features to a program by assembling operations and putting them on a single key. This is tantamount to inventing new products. In some cases, users are remaking the world—for example, deciding that they live in the kind of rational world where the Dvorak keyboard layout won out.

Some macro programs now have additional capabilities for defining menus. The menus can consist of operations already included in the product or of concatenated operations such as showing the list of allowable entries for a field, letting the user choose one, and typing that entry. Again, a totally new feature is added to a program's interface by the user.

The abilities to make new programs act like old programs, to decide what features should be easily accessible, to automate work, and to add features to a program certainly do add to productivity. But they also make users feel in control. The users are deciding how the program should work. They have escaped the control of the programmer and designer.

Macro Facilities Inside Programs

Most major productivity programs, mainly word processors and spreadsheets, include the capability for creating macros. These differ from the separate keyboard macro programs in two ways. First, they often include ways of addressing the capabilities of the program they are designed for, beyond simulating the pressing of keys. Second, they can be created in two ways: either by direct construction in a kind of "macro language" not unlike a programming language, or by "watch me" where the computer records every action the user takes and records it as a macro. The user need only assign the resulting macro to a keyboard, and the computer will do it at any time. This opens the creation of macros to users who cannot or do not care to analyze operations and create commands to carry them out.

Built-in macros have become the hallmark of the sophisticated product. The first incarnation of Lotus JAZZ was panned because it had no macros. Even though it was intended for novice users, it was judged harshly because of the lack of macros.

Sophisticated users make entire applications out of macros, hiding the product's original interface and creating menus or controlled data-entry forms usable by anyone, often by someone much more näive than the macro creator. Here, users actually take on the programmer's role, creating a piece of computer interactivity for someone else to use.

"Work" Menus

Some new products, especially Microsoft Word for the Macintosh, include a menu named "Work." Users can add items from other menus to the Work menu, along with documents and combinations of menu actions not unlike macros. What this does is make what the user wants accessible, accessible. The user, not the programmer or designer, decides what is important enough to be on top. In yet another way, the product can be suited to the user's manner of working.

A lot of these changes just are not that important. While they add convenience, they are not indispensible. Their main function is to put the users in control. Unfortunately, people get dependent on their customized version of the computer, especially with the operating system utilities not discussed in this paper. If one is used to a computer whose keyboard had been redefined, and moves to a machine which does not include that set of keys, increasing personalization does have a backlash.

User-Defined Interface

A few brand-new products, most notably Borland's Quattro spreadsheet, allow the users to define the interface completely, without using macros. Quattro users have three choices: they can accept the Borland set of menu trees, use a menu tree designed to imitate Lotus 1-2-3 closely, or invent a new menu tree. A fourth menu tree is available, designed to simplify things for novices. However, the tasks for novices are different in each office. An advanced user in the same office will invent situation-specific menu trees that make the tasks of that office easier.

Menu trees are created using essentially the same language as is used for macros. Each menu operation and keystroke has a name, and the menus are assembled from names selected by the user. A user can start from any of the three menu trees provided and make small alterations, or start from scratch and take on the whole design task.

The complete user-defined interface does everything that the other user-modification tools do, and more. Users can make an unfamiliar product imitate a familiar one, or add functionality, or decide what is important enough to put at the top level, or what is similar enough to belong together. Users can design their own product without really programming.

But something else is happening, in a purer form than with the other tools. Users who create a menu tree must not only imagine a slightly better way of doing things, must not only long to automate something they already do, but also must analyze what they really do, and how they use a computer. They also must imagine how they would use the computer if they could. Introspection, analysis of their own learning style and working pattern, and a way of imagining the computer as not fixed but their own, all enter into this process. Users become not programmers but either cognitive scientists or teachers. These are probably more comfortable roles because, even though one has to use the analytical skills of a programmer to build an interface using these tools, one does not have to think of oneself as a programmer.

Borland has been advertising a new word processing program called Sprint which also has a user-definable interface. It has been seen by a few people and imagined by a lot more. Word processing is probably the most personal of computer applications, since it is supposed to be invisible. People have strong ideas about what they want in a word processor. They also have passionate attachments to what they are used to. Many offices want to share files but cannot do it easily because everybody is bonded to a different word processor. Maverick employees use their own favorite, and waste lots of time converting their work when the times come to share it. With Sprint, the barrier breaks down immediately. Sprint will come with interfaces emulating the most popular word processors, and probably with a few new interfaces. So the comfort level will increase for people who already love their own word processor, and people who know they can make a better mousetrap are free to try. Control over the computer world is almost complete. The remaining problems are mostly hardware problems, i.e., computers can still break, of course, and they are always too slow.

Hypercard

Hypercard is a program that is all interface. It exists to be an interface to information. Hypercard programs come in the form of stacks, which can be a little bit like databases or a little bit like interactive education, or a lot like front ends to complex information. With Hypercard, the user is always looking at a screen covered with pictures, information, and potential actions. The user takes an action, usually by clicking the mouse someplace, and the stack processes that action.

A major use of Hypercard is to organize information. One can make phone lists, daily appointment logs, or a customized periodic table chart. Hypercard is sold with some generally useful stacks, including an address book and a calendar, which are meant to be altered to suit the user.

Hypercard invites and even demands customization. Even if one uses only the ready-made stacks, additions are needed to make them useful. And changing relatively simple things is incredibly easy. Of course, doing complex things is hard, but the ratio is reasonable—the amount of effort it takes to do something complex is more or less commensurate with the complexity of the task. This is in stark contrast with some traditional programming and macro languages, where simple operations are nearly as difficult to set up as complex ones. In fact, some operations that are ferociously complex to program in many languages are simple in Hypercard. And the complex tasks are done as extensions of simple tasks. Easy tasks should be easy and hard tasks should be possible—that's the hallmark of a useful and flexible tool.

Hypercard makes it so easy to modify programs to one's taste that people are once again tempted to fiddle with their computers in the way that BASIC once made the original personal computer users do. The level of excitement generated by Hypercard among users (as opposed to people who expect to profit from it) is extraordinary. This excitement is happening because Hypercard lets people use their computers the way they always imagined—more like a very smart typewriter and less like HAL.

CONCLUSION

One does not have to love the computer to be an effective user. But the ability to control the computer reduces the fear. All these tools that are now available make it possible and attractive for mere users to control and personalize their computers. The gulf between user and programmer has been reduced. The growth of the nonprogramming power user both made possible and was made possible by the rise of user-interface tools.

REFERENCES

Lammers, S. (1986). Ray Ozzie interview. In *Programmers at work,* pp. 174-189. New York: Microsoft Press.

Turkle, S. (1984). *The second self: Computers and the human spirit.* New York: Simon & Schuster.

MARTIN A. SIEGEL

Associate Professor
Graduate School of Library and Information Science
Assistant Director
Computer-based Education Research Laboratory
Department of Educational Psychology
University of Illinois at Urbana-Champaign

Architectural and Instructional Worlds: Insights for Interface Design

This paper will examine the design of computer/human interfaces from the perspective of two older design professions: architecture and instructional design. Insights can be drawn from these design worlds, lessons learned from their successes and failures. By sharing selected architectural images and instructional strategies, the author will attempt to draw a parallel to interface design.

ARCHITECTURAL WORLDS

The word *architecture* is derived from the Latin and Greek for "master builder." The ancient Roman Vitruvius, whose *Ten Books on Architecture* is the only surviving ancient architectural treatise, gave the term its clearest meaning: Architecture is the union of "firmness, commodity, and delight." It is, in other words, a structural, practical, and visual art. "Without solidity, it is dangerous; without usefulness, it is merely large-scale sculpture; and without beauty . . . , it is not more than utilitarian construction" (Trachtenberg & Hyman, 1986, p. 41). Rob Krier (1988) explained the goals like this: "Architecture has to provide us with physical shelter from our environment, create a framework for our activities and, above all, express symbolic and ethical values" (p. 11).

Function, Construction, and Form

At the heart of all architectural debate is the appropriate balance of function, construction, and form. Of these, function is the starting point for all architectural design. How do people interact with the built environment? What are their needs? Construction is closely related to function. What are structural solutions for a given spatial organization? What materials can be used given the building's climate, landscape, and the availability of natural materials? While these basic questions of function and construction must be addressed by the design, inhabiting the built environment ideally gives enjoyment and aesthetic pleasure as well.

To the ancient Greeks, form was aesthetics. At the Temple of Concord, built in 430 B.C., the Greeks combined mathematics, the "golden rectangle," with construction to create idealized beauty. These were not user-centered structures. The steps, for example, are in proportion with the ideal proportion of the whole, but were out of shape for human scale. These structures were not built for people; they were built for gods.

Unlike the ancient Greeks, more modern architects did not overemphasize form at the expense of function. Nevertheless, these buildings were designed to convey to their users certain attitudes and feelings. King's Chapel in Cambridge, England, and Saint Chapelle in Paris both convey feelings of inspiration and faith. The Bibliotheque Nationale and the Library of Congress Reading Room make patrons feel humbled as they are surrounded by vast quantities of knowledge and information. These buildings do not simply function as structures for the shelter and housing of books; they convey to the patrons how they should think about their environment.

Other structures appear friendly and approachable: a Williamsburg house; a bungalow in Chicago; the Cloud Street Bank in McLeansboro, Illinois; "row houses" in San Francisco; the Cottage Tire Store in Nashville, Tennessee; or Principia College Church in southern Illinois. But what makes these structures friendly and approachable is likely to be a function of our age, our sex, our cultural background, and our notions of what is familiar. The Church of the Sagrada Familia in Barcelona, Spain, is a bit strange and certainly unfamiliar to one unused to viewing a church that looks like a dripping sand castle with tumors!

Other structures conjure up positive images of the past; they are reminders of the classics and the nobility. The New York Public Library is like a Roman temple; the Stinson Public Library in Anna, Illinois, is like an Egyptian or Babylonian tomb; the Old Capitol in Baton Rouge, Louisiana is like a castle; a Hudson, New York house is like a small middle ages castle; and the Trans America building in San Francisco is like an ancient pyramid. Even the conference building at the University of Illinois, the Illini Union, was a 1939 copy of the Wren building on the William and Mary College campus, originally built in 1695.

Some structures, such as the Biltmore House of the Vanderbilts in North Carolina, were designed to express the prestige and wealth of the owner. Other buildings appear to poke fun at the structures after which they are modeled: the Tower of Pizza in New Jersey; the Wigwam Village in Kentucky; the Pagoda Gas Station in Wisconsin; and the Dinosaur Museum in California.

Some structures appear to evoke an overindulgence in form, as does the interior of St. Peter's Basilica. To modern eyes, this structure may seem excessively and unesthetically decorated. But in the sixteenth century, perhaps these overly baroque structures were viewed as an appropriate glorification of God. On the other hand, Mies van der Rohe's glass houses on Lake Shore Drive in Chicago emphasize function over form. These buildings define space, and these structures define their function. Both the Vatican dome in Rome and the glass houses of Chicago upset the delicate balance of function and form.

The French architect Le Corbusier, invoked the "architecture as machine" metaphor. Even his Citrohan House looked like a car. The Bauhaus movement, started by Walter Gropius, promoted the machine-like metaphor not only in buildings (for example, Gropius's apartment building looks like a factory), but in art, drama, and in everyday ways of life. Men and women lived in machine-dominated worlds, and the Bauhaus school promoted the thought that their homes and workplaces should remind them of this.

Le Corbusier expanded this idea as a cultural statement. On paper he designed the "city for three million," a labyrinth of walkways, tunnels, and highrises. His concrete park even accommodated a landing strip for the ultimate machine, the airplane.

However, a smaller version, designed by a different architect, became a low-income housing complex in St. Louis in the early fifties. It was called Pruitt-Igoe, and it was characterized by large apartment complexes surrounded by open spaces. This new urban development did away with traditional streets, gardens, and semi-private spaces. It won an award from the American Institute of Architects. But in 1972, it was blown up by the city authorities. The problem was that the design was totally inappropriate for the needs of the Southern migrants who had no experience living in such densely-packed living compartments. Le Corbusier's "streets in the air" became the site of vandalism, drug abuse, and crime. Pruitt-Igoe did not meet the needs of its users.

Must there be a tradeoff between form and function? Certain

designs appear to be in balance: Frank Lloyd Wright's Fallingwater in Bear Run, Pennsylvania, magically blends into the surrounding environment. Monticello, Thomas Jefferson's home, and the University of Virginia meld form and environment. The Santa Cruz dormitories on the University of California campus allow students to recreate their own individualized spaces; and the Hyatt Regency Hotel in San Francisco creates an outdoor environment indoors.

Instead of looking at a building's form or function, perhaps one should look at its façade. This is what most people experience first: an introduction to a building. Some façades, such as that of the Citicorp Building in New York, give no indication of what is inside. However, the Notre Dame Cathedral façade clearly expresses what happens inside.

Or one can think of façade as entryway. Beyond the entrance to St. Peter's in Rome, and the doors of the Notre Dame Cathedral, it would be surprising to discover an aerobics and exercise center! In a Japanese garden in Kyoto, user perspective is important; the design draws and leads the viewer.

What can be concluded from this brief tour of architectural images? When a building is out-of-balance, the user is dissatisfied. A building may be aesthetically pleasing, yet awkward to use. Or it may be of solid construction and utilitarian, but also dull and uninspiring for its residents. For a building to be successful, there must be a delicate balance among function, construction, and form.

Similar statements can be said for the design of computer/human interfaces. The interface must allow the user to complete the desired tasks effortlessly and transparently (function). The interface must be efficient in execution (construction). And, finally, the interface must be pleasing to the user (form). It is difficult to measure the negative effects of misuse of color, typography, and graphics; but all play a part in the level of user satisfaction with the interface.

Examples of Interface Design

This paper will examine the entryways to two different computer systems. The first is the logon procedure for the UNIX system at the University of Illinois (see Figure 1). The date and time are written in an unfamiliar format (year/month/day and twenty-four-hour clock). The user types "name" as "231004321" and must indicate who will pay the computer time bill.

The second entryway is the logon procedure for the NovaNET system at the University of Illinois (see Figure 2). The date and time are displayed in familiar formats, and the system greets the user with a "welcome" message. The user types his or her "NovaNET name" as

```
TERMINAL:  357
88/04/18.  11.05.18
UNIVERSITY OF ILLINOIS ALL ACCOUNTS   ::NOS 1.4  501/498
SIGNON:  231004321
PASSWORD:
TERMINAL:  357,TTY
RECOVER/CHARGE: bill,psych,ps1204
```

Figure 1

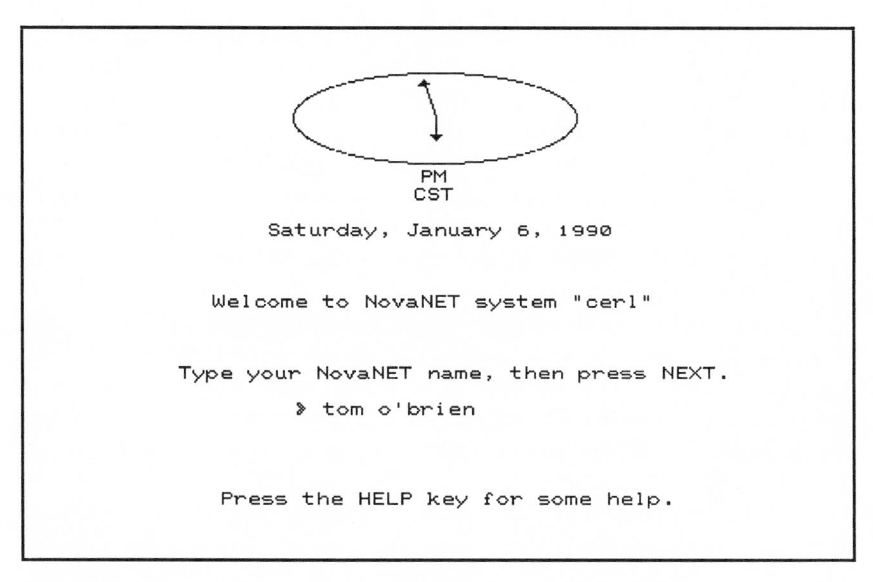

Figure 2

given by the instructor: tom o'brien. On the screen, the user is prompted to press the HELP key (an actual key on the NovaNET keyboard) if assistance is required.

To be fair, the first screen was designed for computer scientists and students of computer science; the second screen was designed for students and instructors enrolled in a computer-based curriculum. Nevertheless, the NovaNET entryway is accessible to all users. The interface's image communicates an important message to its users: This system is for you; it will be easy to use. The UNIX interface image communicates a different message: You'd better know what you're doing if you're going to use this high-powered system.

Architects sometimes use metaphor to communicate meaning. For example, Le Corbusier's Citrohan House glorifies that favorite machine, the automobile. "The house was to be a mass producible 'machine for living' to alleviate the severe postwar housing shortage. . . . The model projected an aura of salutary efficiency and vitality. What is more, it was named the Citrohan House not only by analogy, for if one steps back and looks at the 1922 model, it has the generalized shape of a 1920 Citroen sedan" (Trachtenberg Hyman, 1986, p. 529).

Computer/human interface designers use metaphor as well. For example, the Macintosh interface makes extensive use of the desktop metaphor. Familiar objects such as the trash can, clock, Rolodex, and bookshelf are used as icons and "interface façades"; files are opened and appear as sheets of paper overlaying other sheets of paper (see Figures 3-5). Is there any doubt that when one drags a file to the trash can one is destroying the file? The interface takes advantage of a person's common knowledge to increase ease of use.

Compare these interfaces with the Disk Operating System (DOS) used on IBM machines (see Figure 6). Command statements such as "A>diskcopy a: b:" remind the user that the machine must be engaged on the machine's terms. Le Corbusier would have made a good C.S. major!

Other computer/interface designs successfully balance the architectural equivalent of function, construction, and form. Two popular examples are *Microsoft Word* (Microsoft Corporation, 1989) and *SuperPaint* (Snider et al., 1986) (see Figures 7 and 8). *Word* makes use of the

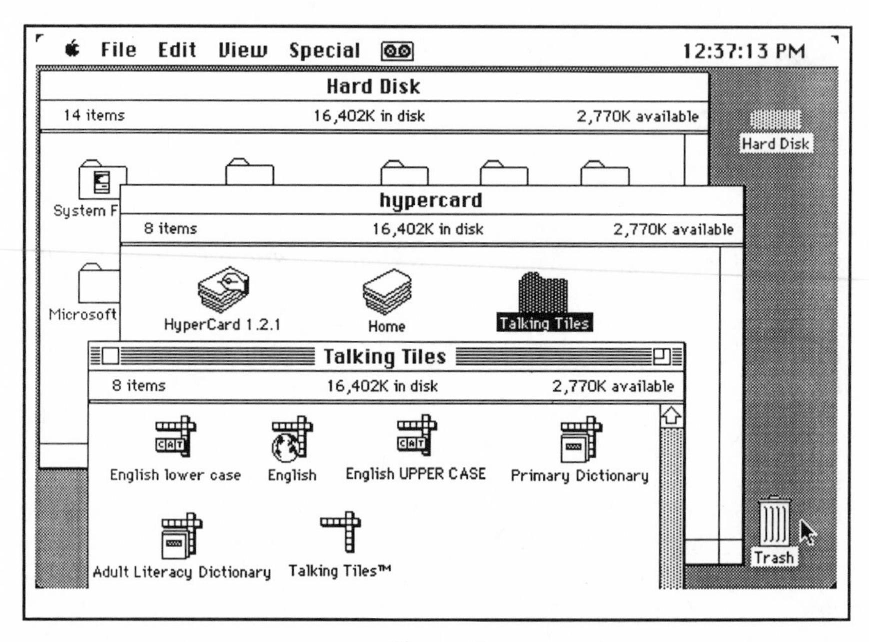

Figure 3

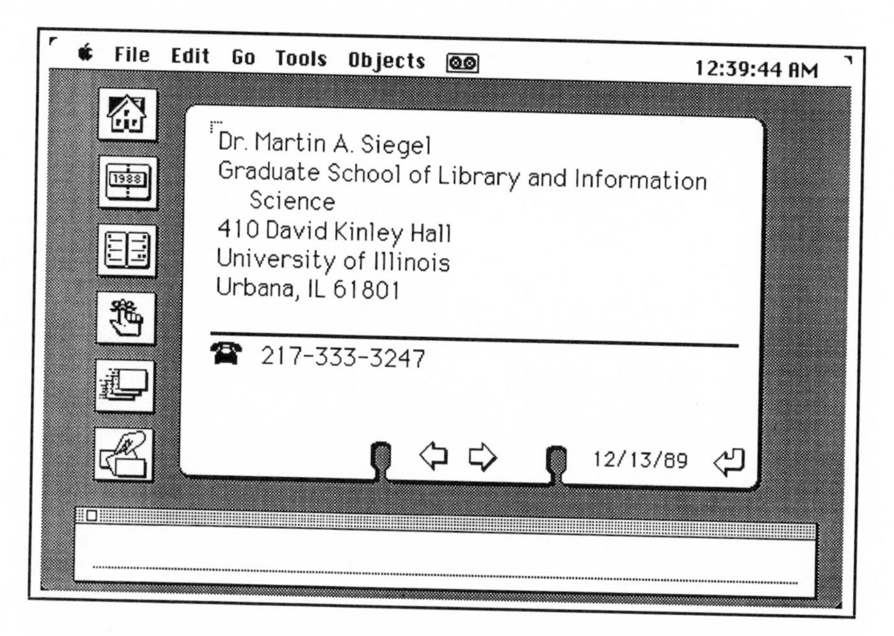

Figure 4

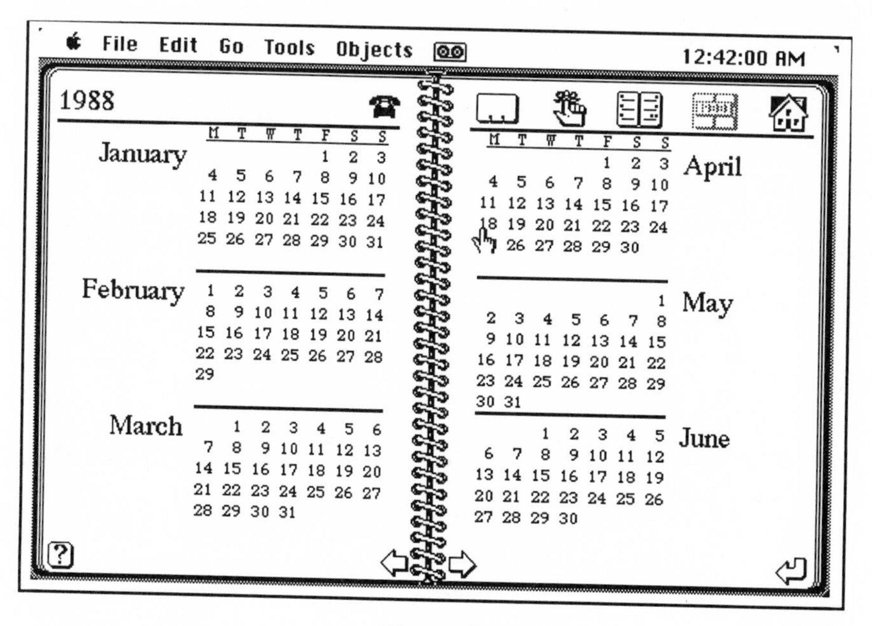

Figure 5

```
Current date is Tue  1-01-1980
Enter new date (mm-dd-yy):
Current time is  0:00:07.79
Enter new time:

The IBM Personal Computer DOS
Version 3.10    (c) Copyright IBM 1985
                (c) Copyright Microsoft Corp 1985

A>format b:
Insert new diskette for drive B:
and strike ENTER when ready

Formatting...Format complete

     362496 bytes total disk space
     362496 bytes available on disk
Format another (Y/N)?n

A>diskcopy a: b:
```

Figure 6

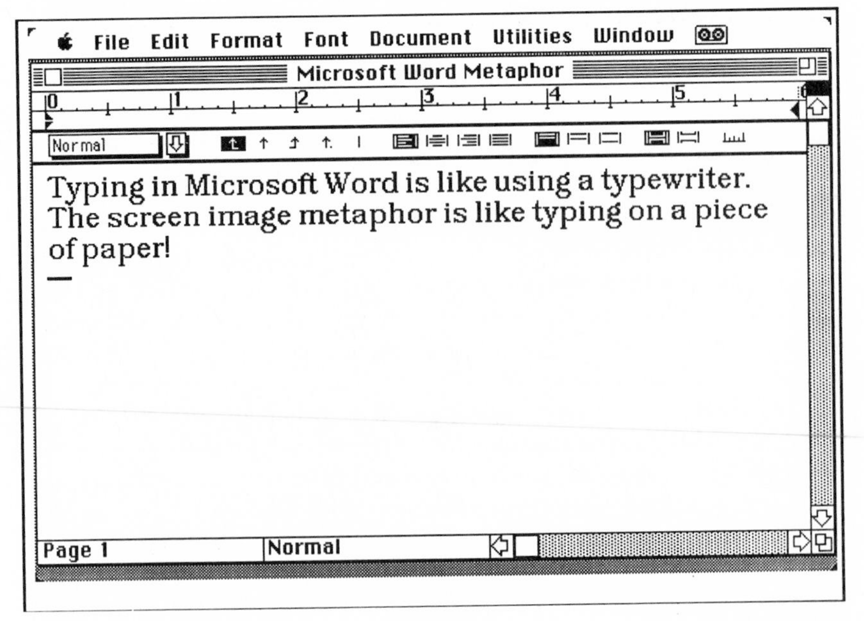

Figure 7

typewriter metaphor to enter text as it would be printed on paper. It was the first widely distributed word processor to give a true "what you see is what you get" feeling. *SuperPaint* allows use of paint brushes or spray cans to "paint on walls." Pictures and graphics are produced in powerful ways, creating a new art form.

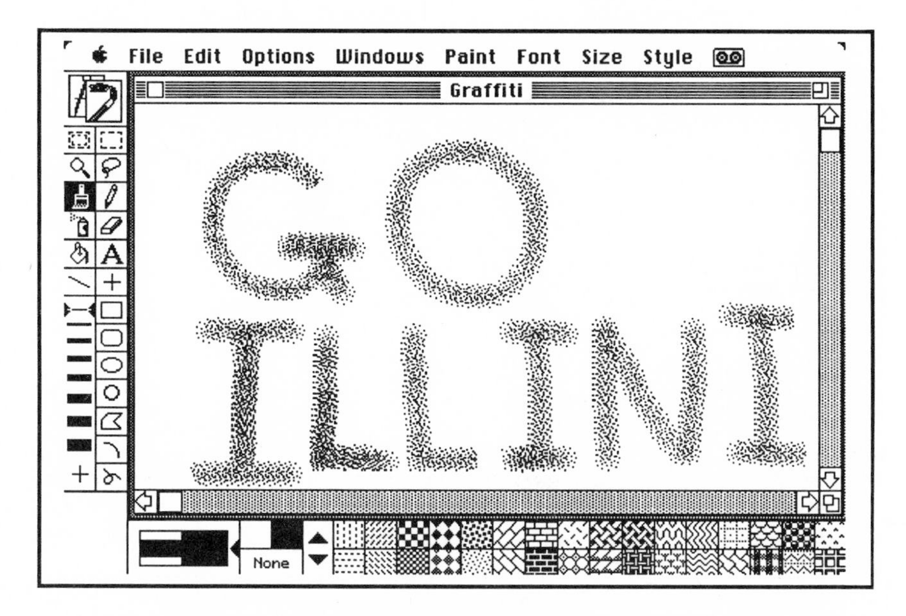

Figure 8

INSTRUCTIONAL WORLDS

Instructional design is at the intersection of three separate analyses: the analysis of behavior, the analysis of communications, and the analysis of knowledge. The analysis of behavior seeks empirically based principles that provide the basis for teaching any task: how to motivate and get attention, how to present examples, how to secure student responses, how to reinforce appropriate answers and to correct mistakes. This part of the analysis can be summarized as the mastery-learning model.

The analysis of communications seeks principles for the logical design of teaching sequences that effectively transmit knowledge, prevent the learning of misrules, prevent overgeneralization to inappropriate examples, and deter undergeneralization. This analysis focuses on the ways in which sets of stimuli are the *same* and how they are *different* (that is, on what discriminations must be taught).

The analysis of knowledge systems is concerned with identifying commonalities across different concepts or procedures. This provides a basis for teaching broad general cases, "big ideas" or insights, that provide a framework for understanding. Moreover, similar knowledge structures share similar communication strategies; that is, if two sets of concepts are similarly structured, then they can be taught in similar ways. It is these last two analyses, communications and knowledge, that most influence the design of computer/human interfaces.

The new user of a computer program is faced with the task of learning a set of concepts and procedures associated with that system. Learning is accomplished primarily by what is communicated to the user by the interface. In designing an interface to best facilitate that learning, designers must understand how generalizations are learned. Every interaction with the user is a potential opportunity to learn a generalization. Sometimes, however, an incorrect generalization or mis-rule is learned.

Teaching Generalizations Through Interface

What is the single greatest advantage for learning a generalization? Efficiency. Indeed, without generalizations we would not know how to function. Fortunately, humans are "wired" for generalization learning. When a baby hears the word *daddy* and sees a particular man, the baby begins to make a connection between the label and the person. Sometimes the man is wearing a suit, sometimes he is wearing pajamas, but *that man* is always daddy. Over time, the generalization is formed.

Babies have a "natural" way of learning. Direct instruction is minimal. No one has to teach babies that when they turn over a pail of water, the water pours out. In this case, gravity is a good teacher. Moreover, babies have plenty of time to learn the principles of carrying water in pails.

The adult learner has no such luxury. The boss hands the employee a copy of the company's chosen word processor and says, "Learn it." The boss expects the employee to master the computer tool within days rather than years. Learning is more cognitive than physical. No matter how many times the user forgets to press a particular function key to mark text, the environment does not provide the kind of feedback that it does for babies carrying pails of water upside down; gravity does not affect word processors. The computer/human interface designer, therefore, cannot depend on "natural" and casual learning. The designer must create an interface that is immediately understandable and easy to use so that the employee can master the product quickly.

The challenge for the instructional designer is to analyze a concept's features so that generalizations can be taught. Similarly for the interface designer, the challenge is to analyze the product and design the product's interfaces so that the user can take advantage of any underlying generalizations. If every command is presented as a unique case, the learning rate is slower than if clusters of commands are taught as a single generalization. To say this another way, the user either thinks about many boxes of concepts or procedures, or the user thinks about one unified box. Obviously, learning a single unified concept is more efficient than learning many seemingly unrelated concepts.

But what box? The trick is to identify the most powerful boxes—those that give the user insight into using the tool. To take a simple example, can all menus in the system be opened in the same way? Does the interface make this clear? Big concept teaching empowers the user.

Examples of Interface Design

If users only need to insert text from the beginning straight through to the end, word processors are easy to use. But writers must edit their work; they must return to the text frequently to make additions, deletions, and corrections, and to rearrange a draft's content. When using paper and pencil, this operation is relatively easy. The writer merely picks up a pencil, moves over the page, erases or crosses something out, and inserts new material.

Of course, if the writer makes extensive changes, the draft becomes difficult to read, and if the writer wants to move blocks of the text, the pieces are cut and pasted. One of the word processor's chief advantages is that, when the writer makes these changes, a "clean copy" can be printed incorporating the modifications. Nevertheless, with paper and pencil these operations are at least conceptually easy to grasp and thus transparent. They are not so conceptually easy on a word processor. Before a user can actually make the desired changes in a text, the user must tell the computer where the changes will occur. To do that, the user moves a cursor, or indicator, over the screen until it appears in the desired place within the text. Then the user can add, delete, or move things from that point. But the user must first learn to use a series of directives that control the movement of the cursor, and that is where the transparency problem occurs. If writers must learn and use a large number of commands simply to move the cursor on the screen, they can easily be distracted from writing and thinking. This does not occur if they are using paper and pencil. Yet, if the word processor is to provide writers with power and flexibility in writing, many such directives must be available.

A few years ago, the author and a colleague designed a new educational word processing system called *Electronic Ink* (Siegel & Felty, 1986). The system solves the cursor command dilemma by employing generalizations that govern keypress conventions. One such generalization governs "amount of move." For example, in Figure 9, the arrow keys move the cursor horizontally, while the Ctrl and Alt keys add power to keypresses. Thus, the left arrow moves the cursor one space to the left; Ctrl-left arrow moves the cursor one word to the left; and

Alt-left arrow moves the cursor all the way left, to the first character on the line.

```
     Left arrow = moves cursor one space to the left
Ctrl-Left arrow = moves cursor one word  to the left
 Alt-Left arrow = moves cursor one line  to the left

     Right arrow = moves cursor one space to the right
Ctrl-Right arrow = moves cursor one word  to the right
 Alt-Right arrow = moves cursor one line  to the right
```

Figure 9

The generalization is:

Key = small move
Ctrl-key = medium-sized move
Alt-key = big move

Then, when students learn to use the backspace directive (Bksp) to erase text to the left of the cursor, they employ the same generalization:

Pressing Bksp erases one character
to the left of the cursor.

Pressing Ctrl and Bksp erases one word
to the left of the cursor.

Pressing Alt and Bksp erases everything on that line
to the left of the cursor.

The same relationship exists between the right arrow and the delete directive (Del), which erases text from the cursor to the right. Moreover, the directives Home and End utilize the same generalization. Pressing the Home key takes the student to the first character on the screen; the End key takes the student to the last character on the screen. Using the Ctrl key with Home and End gives bigger moves—to the first and last characters in the file.

Everything about an interactive system is communicated to the user through the interface: the keypresses, the procedures, the overall structure, and even the inconsistencies. While this presents many problems to the computer/human interface designer, it presents opportunities as well. Knowledge about instructional design principles can be used to facilitate good interface design.

CONCLUSION

The world of architectural design reminds us that interfaces are more than art objects. They are designed to interact with people. They must be constructed in such a way so as to balance function and form. The interface of any system must do what it is intended to do, but it must do so with flexibility appropriate to the user's needs. The best interface design helps the user think about tasks in powerful ways.

The instructional design world serves as a reminder that people learn from their environment and that the most subtle cue may trigger learning (misrules as well as rules). A well-designed interface can make learning easier by taking advantage of people's ability to generalize. Similar procedures should be implemented in similar ways. A powerful interface makes visible a product's features. The user is freed to concentrate on the task at hand rather than on the idiosyncrasies of the machine.

REFERENCES

Krier, R. (1988). *Architectural composition.* New York: Rizzoli International Publications, Inc.

Microsoft Corporation. (1989). *Microsoft word: Document processing program version 4.0 for the Apple® Macintosh™.* Redmond, WA: Microsoft Corp.

Siegel, M. A., & Felty, J. M. (1986). *Electronic ink: An educational word-processing system.* Chicago, IL: Science Research Associates, Inc.

Snider, B.; Zocher, E.; & Jackson, C. (1986). *SuperPaint.* San Diego, CA: Silicon Beach Software, Inc.

Trachtenberg, M., & Hyman, I. (1986). *Architecture: From prehistory to post-modernism / The western tradition.* New York, NY: Harry N. Abrams, Inc.

WALT CRAWFORD

Principal Analyst for Special Projects
The Research Libraries Group, Inc.
Mountain View, California

The Online Catalog in the Real World: Making the Necessary Compromises

INTRODUCTION

Online catalogs have been in existence for several years now, but they are still in their infancy as compared to other aspects of library automation and librarianship. Today's best online catalogs are more powerful, more flexible and easier to approach than were most early examples; tomorrow's best online catalogs should be better than today's. Two characteristics have always been true of online catalogs and will always be true: first, that there will be several distinctly different good designs; second, that every good design will reflect conscious choices among different desirable features, some of those choices involving compromises.

In a very real sense, and unlike many other computer programs, an online catalog is nothing but an interface. A library catalog links library patrons with library collections. A good library catalog brings readers together with materials they need and with materials they can use but were not aware of. An online catalog is a view of the bibliographic database; it is also an interface between the patron and the collection. This paper is really about choices: the choices needed to make catalogs work in the real world. Some choices do require compromise, not necessarily as signs of failure or as short-term necessities. While unconscious choices may lead to poor systems, it is not possible to create a good online catalog without making choices and compromises—and intelligent conscious choices will yield the most useful catalogs.

Most librarians have abandoned the idea that there is a perfect, paradigmatic design for an online catalog. There are still those who believe that there is a single ideal design for online catalogs, applicable to all institutions and all users, and that research and development will determine that design which could then, presumably, be embodied in a set of national standards. It is hard to accept this view, since no single catalog will satisfy all patron access needs for all patrons in all libraries.

EXAMPLES

Some of the following examples will illustrate cases in which a catalog designer must make choices, deliberately or accidentally.

Menus or Commands

Should an online catalog use menus or commands? The online catalogs should allow patrons to move directly from one type of search to another, and patrons should generally be able to take any legitimate action at any point within an online catalog interface. At the same time, first-time users should be able to use an online catalog with little or no instruction, preferably without needing to ask for help.

If properly implemented, command-driven systems provide direct access to and movement from one point to another; the commands should always have distinct meanings. Menu systems, however, help the first-time user to perform a search by offering descending levels of choices. When a designer chooses to use numbered menus (or their equivalent, using movable light bars), the designer essentially chooses not to support direct movement within a system—the two are largely incompatible. When a designer chooses to use commands, first-time users are much more likely to require some help.

One Interface or Multiple Interfaces

Should an online catalog have only one mode of operation, or should it work differently for different patrons? Although the catalog interface should suit the skill and needs of the patron, the patron should also move along the learning curve rapidly, using the catalog more effectively each time he or she searches.

Some catalogs include multiple interfaces to try to suit the interface to the user. For example, a catalog might have a pure command-driven mode and a mode that is so heavily prompted as to be essentially menu-driven; it might even have a conversational mode. But catalogs with multiple interfaces don't allow easy movement along the learning curve. When patrons have completely learned one interface, they're at ground zero with the more "expert" interface; the learning curve turns into a series of cliffs. Worse yet, they must decide at the beginning of each session whether they are novice or expert and must know enough to make an intelligent decision. Patrons familiar with one interface who sit down at terminals abandoned in mid-session using a different interface may be totally confused by what's on the screen and uncertain about how to proceed.

Library Customization and Maintainability

Should online catalog interfaces be customized to the needs of each library and its branches, or should every installation of a given system be identical? On the one hand, each library (and possibly even each of its branches) should be able to choose the indexes it requires, the text to be used in prompts, messages and help screens, and the set of functions it will support; for a menu-driven system, it should even be able to choose the arrangement of menus. That allows each library to suit its own patrons and policies. On the other hand, patrons who use more than one library should be able to make use of their familiarity with one catalog while using catalogs in other libraries, particularly if the catalogs appear to be identical or similar.

If each library has a different version of a system, patrons may be betrayed by their own familiarity, and vendors will need to spend time determining whether problems are part of the underlying system or part of a given library's implementation. For that matter, enhancements and extensions must fit into each library's version, possibly requiring rethinking for some or all of the implementations. Flexibility, in this case, conflicts directly with maintainability. This problem is a serious one, not generally acknowledged; those agencies who supply many libraries with catalogs face serious maintenance and upgrading problems that directly affect the libraries.

Labeled and Unlabeled Displays

Should the data elements displayed as the result of a search be labeled and appear on separate lines, or should bibliographic fields be presented in compact displays similar to catalog cards? Providing labels for bibliographic elements makes displays easier to understand, but patrons can deal with single records more readily if a complete record fits on a single screen. Some patrons need all available information on a record, and all patrons can benefit from relatively complete information. Displays should not be cluttered, and generally should not be more than 15 to 30 percent full—that is, 300 to 600 characters in a typical display.

All of those statements are well-supported by various studies in various fields, but they set up a series of internal conflicts for catalog design. Labels take up space, and, more specifically, require more lines to display a given amount of information. When the Research Libraries Group (RLG) studied medium-level displays, including all access points (subjects and added entries) but leaving out notes, 97 percent of the records in a very large sample could each be displayed in a single screen in an unlabeled format—but only 41 percent of the sample would each fit on a single screen with labels (Crawford et al., 1986). That result argues strongly for an unlabeled "card-like" display. On the other hand, the consensus is that patrons will find labeled displays not only easier to use but also more meaningful.

Speed, Versatility, and Ease of Searching

Which is more important for library patrons (and, thus, well-designed online catalogs): finding known items rapidly, being able to find items on a subject without knowing the library's name for the subject, or being able to search for material easily if not rapidly?

On the one hand, patrons should be able to find known items, their call number, and their availability immediately. Additionally, experienced patrons should be able to use advanced search techniques to locate specific items from very large databases. On the other hand, a catalog should also help patrons to find items when they aren't quite sure what the item is and it should help guide them to material on a given subject, even if they don't think of the subject the way the Library of Congress does.

Fast Boolean searching using a command language, with immediate record display for single results and small results, almost certainly offers the best performance for known-item searching. Boolean searching is fundamental to advanced searching techniques and almost mandatory for very large databases, but Boolean searching may do more to confound new users and 'computerphobes' than add to the efficiency of a catalog interface. Searching authority lists of authors or subjects, or browsing tables of call numbers, helps patrons to find items when they're not sure what they want, and helps them locate all the works they may need but inevitably slows down known-item searching. Offering fast known-item searching, authority-list browsing, and complete Boolean capabilities may result in a complex, difficult-to-learn interface.

There have been conferences and books dealing with nothing more than search techniques and subject retrieval, and they have neither resolved all the issues nor uncovered all the possibilities.

Anonymity and Extended Functionality

What should the catalog know about the patron? An online catalog will be much more useful if the patron can place holds on material, issue requests for interlibrary loan (ILL) when the catalog includes holdings for other libraries, suggest orders for materials not found, and send messages to the library. Additionally, a good case can be made that the catalog should adjust its own performance to the specific habits and needs of each patron.

Patrons should be able to start using a catalog immediately, be able to leave a catalog without going through any extended sign-off ritual, and be confident that their searching behavior is private—that neither the library nor any outside agency knows their personal searching behavior. However, the catalog can't support holds, ILL requests, or order suggestions if it doesn't know who the patron is; it cannot adjust its own performance unless it maintains records on the patron. But if a patron must identify himself or herself, it will take longer to start a session, will create problems if the patron doesn't explicitly end the session, and will certainly weaken the patron's confidence that the library isn't keeping track of what he or she does.

Clean Screen and Full Information

Finally, how much and what kind of information should appear on the screen at any one time? On the one hand, people can apparently recognize and deal with five to nine options more rapidly than with a larger set of options. People can generally make choices more readily from visible options than from remembered options, especially when they are just learning a system. On the other hand, patrons can concentrate better on the information at hand if there is little or no other information on the screen, and people can cope most readily with screens that are only 15 to 30 percent full, and with text that includes no more than about 60 characters to a line.

Once again, multiple desirable traits come into conflict, particularly given some other very desirable aspects of an online catalog (such as flexible searching and the freedom to move from anywhere to anywhere else). A full-featured online catalog has more than nine reasonable choices at several points, and always has more than nine possible actions. For that matter, displaying nine options, or even five or six options, will add significantly to the density of a screen and to its complexity.

A similar quandary arises in presenting online help and online tutorials. A good message fits on a single screen and is no more than 600 characters but a good message also explains the situation completely.

It is difficult to explain Boolean searches in 600 characters; it is even hard to explain truncation clearly in a hundred words or less.

BACKGROUND

The Research Libraries Group is a consortium of major universities and research libraries that pursues common aims in library and scholarly fields. The most visible aspect of RLG is RLIN, the Research Libraries Information Network, which serves as the computer support for all RLG activities and provides shared technical processing support over a national telecommunications network linking more than a thousand terminals. RLIN provides one of the most powerful, sophisticated, and flexible bibliographic retrieval systems available today.

What RLIN does not provide is a sophisticated user interface. The system was designed for use by library staff who have received some training; it was not designed for direct use by scholars or other patrons (although it is now being used directly in some cases).

In 1984, the J. Paul Getty Trust funded a two-year RLG project with a number of aims. One of the aims was to develop a design for a workstation-based patron access system, to work with an online catalog based on RLIN software. It was felt that the RLIN software could provide exceptionally good support for the kind of online catalog needed by scholars, and that a proper user interface would be needed to make such a catalog worthwhile.

The first phase of the project was to investigate the literature of patron access and develop a documented sense of what was being done and being suggested. The author took up that task during 1984 and most of 1985, going somewhat beyond the original charge to develop an overall outline raising more than 250 specific issues relating to patron access. Midway through the project, the author attended a conference on online catalog screen displays, sponsored by the Council on Library Resources (CLR) which was a source of motivation to write about patron access issues, and to develop a software system that would allow RLG to run large-scale tests on various screen designs (Crawford et al., 1986), and to further develop the issues outline (Crawford, 1987).

Most of the content of this paper is based on the author's experience in designing an interactive retrieval system with an online patron access catalog, and working with representative intended users of that catalog. A careful analysis of all existing online catalogs has not been carried out in this paper, the aim is not to focus on specific systems in use at present. This field is growing rapidly and the designs of online catalogs change quickly too, making it hard to document them.

Recognition of Mutually Exclusive Desirable Features

The book *Patron Access* (Crawford, 1987) includes many assertions as to desirable features and aspects of an online catalog but does not constitute a prescription for an ideal one. Indeed, combining all of the desirable features into a single catalog, if that is even possible, would probably yield very poor results.

A good human/computer interface is not simply a collection of good features; it is a coherent system that makes overall sense to the user. In order to build a coherent system, a designer must make choices; some of those choices will eliminate features that might (in the abstract) be considered desirable. Some desirable features exclude other desirable features. Following are some categories of choices and compromises, which are somewhat arbitrary and definitely overlapping.

ECONOMIC COMPROMISES

Economic compromises lead system designers into easy traps, and to the assertion that "if only we had enough computing power or an adequate database, we could do everything we'd like to." Then, when a library can afford ten times as much computing power and has the kind of database systems designers said it should have, those same designers are bound to admit that the system never will be able to do "everything we'd like to." As economic limits are being removed, analysts and vendors are placed in a somewhat perilously exposed position. Following are a few examples of compromises imposed by economic limitations.

Computers with Insufficient Power

Online catalogs require powerful computer support. In the past, libraries have rarely been able to afford computers powerful enough to mount the level of catalog support they really wanted. When the computer lacks sufficient power, a designer may need to choose between offering Boolean searching (typically a heavy strain on computing power) and crisp response. Similarly, it takes computing power to sort the results of a search; some designers have chosen to present results randomly in order to reduce the load on the computer.

Some design decisions seem to mask inadequate computing capacity, although they may not have been made for those reasons. Perhaps the most extreme example of this was the touch-screen catalog that effectively used the patron as the balanced-tree index. Since the patron saw a screen (and had to take action) for each level of the index, very little computing power was required between screens, thus making good response available from underpowered computers. Of course, it took a long time to actually find anything, but patrons didn't have to wait very long between screens.

Less extreme examples, probably not consciously designed to mask computer inadequacies, are the strict menu-driven systems in which a patron can only choose from a list of numbered options on each screen. To get from one kind of search to another, the patron may need to back up through the screens to the top level, or restart the catalog session to get straight to that level. That design compromises speed and fluidity of use. Designers would probably say that the compromise was to favor ease of first-time use. At the same time, the design can help to mask computer problems in two ways: first, a strict menu-driven system eliminates parsing, and parsing places some load on the computer; second, and more significantly, strict menu-driven systems rarely allow effective interindex Boolean searching, and generally don't allow for searches that strain the abilities of a system.

Pure menu-driven systems may be a reasonable compromise between computer power and patron effectiveness for miniature microcomputer based catalogs designed for very small libraries. Most of the other pure menu-driven catalogs perhaps do not make that choice as a conscious attempt to save computer power. Pure menu-driven designs save programming and analysis even more than they save computer power. It is easier to design and implement a set of menu trees than to design and implement a command-driven system in which a patron can go directly from one function to another. The problem is that such designs compromise the patron's long-term speed and effectiveness for the sake of easy implementation.

Problems with computer power may always exist, but powerful computers get cheaper every year, and cheap computers get more powerful at a rate that almost defies belief. Today's Atari ST computer at less than $500 is, in terms of raw processing power, more powerful than many minicomputers of 1980 and some mainframes of 1975—and Atari is introducing "transputers," very powerful and inexpensive boxes that can be connected to provide almost unlimited processing power at very low prices. Enormously powerful cheap computers don't, in and of themselves, provide the means to eliminate computing power as a reason for compromises in online catalogs, but they are a step in that direction.

A serious assertion was made, not too many years ago, that it wasn't reasonable to build an online catalog for a collection of more than half a million titles. There could never be enough computing and indexing power to make online retrieval work in that large an environment—in effect, half a million titles was a reasonable upper limit for a good design. This assertion, however, was not true, since at the time, RLIN was providing good online retrieval to 2 or 3 million titles, to name one counterexample. Today, MELVYL (University of California Online Union Catalog), a MARC database of documents held by University of California libraries includes more than 3 million titles, and RLIN offers fast access to nearly 30 million records and at least 8 million titles. Any number of universities have online catalogs with million-plus databases.

Admittedly, RLIN runs on a very large computer, as does the University of Illinois catalog, UCLA's ORION, University of California's MELVYL, and most other large campus catalogs. On the other hand, University of California at Berkeley's GLADYS runs on what is called a minicomputer, and very large catalogs may now be feasible based on microcomputer architecture. What is left is the formidable task of designing catalogs to take advantage of that cheap power, finding the analysts and programmers to implement the catalogs, and making them work in a coherent fashion.

Storage Subsystems with Insufficient Capacity/Speed

The biggest reason for compromising the power of online catalogs has been computer power, but problems with disc storage have run a close second in some cases. Complete bibliographic records take up a lot of disc space, comprehensive indexing takes up even more space, and sophisticated searching may involve manipulating a large number of records. That all means large disc storage requirements and the need for high-speed disc access. Such capacity and speed have not come cheaply.

The cost of disc space has led to two compromises in catalogs and other library systems, both of them unfortunate. The first compromise is to store only "needed" information in the database, throwing away portions of the MARC records or other information sources. The second compromise is to limit indexing and retrieval. Of the two, the first is the more problematic in the long run. It is possible to add indexes to a well-designed system as discs become cheaper, larger, and faster, but there is no sure way to restore lost data.

Discs have become cheaper and faster, although not at the remarkable rate of computers. It is still true that small computers have a great deal of trouble handling very large disc systems, but those problems will pass. While high-speed hard-disc storage may never become as cheap as high-speed computing power, discs continue to become more affordable, and libraries can eliminate compromises required to save disc space.

Installations with Insufficient Terminals

For years, accepted wisdom was that no online catalog ever had enough terminals. That is not true anymore, but it has historically been one of the major complaints. Curiously, design decisions don't seem to reflect the compromise that one would expect to see when there may not be enough terminals—catalogs designed to increase access to information on a known item, and to do call number searching quickly and easily. Those features would cut the search time of many users, allowing a limited number of terminals to serve more users. Fast known-item searching—and direct searching in general—is fairly rare in public library oriented online catalogs. That suggests that a probable shortage of terminals simply hasn't entered into the design process.

One way to deal with a shortage of terminals is to help patrons use them effectively. Sometimes, the best way to do that is to provide information through other means than the online catalog. A prime example in most public libraries and many academic libraries is subject browsing by patrons who really want to browse through part of the collection and aren't sure which part of the stacks they need.

Browsing the stacks is a perfectly legitimate activity, one that libraries should generally encourage. No bibliographic description will ever be perfect, the description of a book is not the book itself, and only the user can determine what will really meet his or her needs or wants. Patrons wanting to browse don't really want to spend time with the online catalog; they aren't as interested in the status of a given item as in the range of material available, and they want to inspect the material. They could certainly get some call numbers using the online catalog, but that is a fairly extended process in many online catalogs involving four to five steps or more.

One local public library helped these patrons by preparing a paper list, based on the thousand-division Dewey table, expanded and sorted alphabetically by topic, short enough for easy use and long enough to offer a wide range of possibilities. It was checked against the actual collection so there are no blind references: every heading in the list reflects at least one title in the collection. The list was an experiment; a few copies were placed near terminal clusters. The reactions were positive. Some patrons wanted to buy copies of the list, the copies were heavily used, and more copies were printed and made available. In this case, the list was not a way of freeing up terminals but a way of serving patrons who really didn't feel the need to use terminals. But the principle can serve both ends; the online catalog can be more effective if it is *not* the only source of information about the collection.

Incomplete Databases: The Catalog is the Collection

The most agonizing compromises in online catalogs are those required because the catalog database is incomplete or inadequate. When half the titles in a collection are represented only by truncated portions of the author and title, searching performance can never be wholly satisfactory. Methodologies that will provide the best performance on good quality records may fail to identify brief records; access by subject or by other fields may be wholly impossible or may give an unrealistic picture of the collection. When major portions of the collection are not reflected in the catalog, the catalog undermines the collection.

This problem is more severe in large libraries than in small and, unfortunately, it is large libraries that are generally least capable of preparing complete databases consisting of complete records, at least without spending massive amounts of time and money. In a small or medium-sized public library, people will tend to explore the stacks, as noted above; items not reflected in the catalog will, in some cases, be found serendipitously. Of course, some students and faculty explore the stacks in large academic libraries, but that's a formidable task, particularly when the stacks are likely to be split into two or three dozen different physical locations and may be split by changes in call number policies over time. If a library is large enough that browsing the shelves is not the most practical way to do subject searching, then it is fair to say that the catalog *is* the collection. Materials not fully represented in the catalog are partially discarded because patrons do not have full access to the materials.

Economic Compromises in Current Systems

Every online catalog currently in use reflects compromises made for some of the reasons just mentioned. It will be years before these concerns become secondary, if indeed they ever do. Computers get faster, but librarians and designers continue to devise new ideas that will use even more computer power, and successful projects to complete databases and upgrade partial records will use more disc space and more computer power. In some cases, economic insufficiency of one sort or another has been an excuse for inadequate design; that should be less true as time goes on. But the primary reason to make compromises in catalog design may always be the lack of unlimited resources.

CONTEMPORARY CHOICES

There are some choices required to design a catalog regardless of economic questions. Many choices in catalog design are made unconsciously. That is not to say that they are not the right choices, only that the designers may not be aware that they are trading off some desirable aspects for others. That is less true now than a decade ago; most contemporary designers should have access to fairly extensive literature on catalog design possibilities, the results of user studies, and human interface considerations in other fields.

Simplicity vs. Power

The most common compromise in catalog design, and more particularly in the interface, is to trade power for simplicity and speed of access for presumed ease of first use. These compromises may be reasonable and necessary for a given environment. They may also cripple the power of an online catalog because of needless restrictions on access.

Many online catalogs use menus which are generally presumed to make an interface easier to learn and use. But that presumption has never been thoroughly tested. The Council on Library Resources' study of online catalogs found that menu-driven systems were neither easier to use nor as effective as command-driven systems. But that is not the issue here; the issue is the unconscious choices that arise from a menu-driven interface. There are conscious choices, obviously. For example, a consistently menu-driven interface inherently requires more cumbersome means for building multi-index searches than a command-driven system but might potentially offer a more readily understandable means of building Boolean searches than a typical command language.

But there are also unconscious and possibly needless compromises. The presumed ease of use of menus arises from constant display of options and the ability to take actions without very many keystrokes. Those advantages may require a displayed set of options available at a single keystroke. That doesn't necessarily mean numbered choices and inflexible trees of narrower and narrower menus, however, even though that is the typical implementation. At least one online catalog is presented as menu-driven but is in fact a terse command-driven system using single-character commands that appear as menu options. That is a fairly common methodology in microcomputer software which, if carefully implemented, can offer much of the freedom of a fully command-driven interface with the initial ease of a normal menu system.

Freedom of movement is the first compromise made by most menu systems. Any good command-driven system will permit a patron to go to anywhere from anywhere as long as the transition makes sense. Thus, if a patron is looking at a record display that results from a title search and recognizes that he or she wants more information on a certain subject, the patron should be able to move directly to a subject browse or search, without escaping from the record display, back through the title search screen, back to a screen offering a choice of indexes. Most command-driven systems allow free movement; very few menu-driven systems do. It is inherently impossible to allow free movement in a system that works only with numbered choices; for one thing, the numbers have no meaning except within the context of a certain screen.

Most designers of menu-driven systems do not see the lack of free movement as a compromise; they are sold on the obvious ease of stepping down through menus. But it doesn't take patrons long to recognize the value of direct access. The inability to move directly from one function to another becomes an ongoing irritant, perhaps not major but certainly not necessary. A terse command-driven system can, with careful design, permit relatively direct access while maintaining heavy prompting. It requires clever design and implementation but is at least theoretically possible.

Index Complexity and Versatility

Another set of choices and compromises involves the set of indexes used in a catalog. Every choice has consequences, and every choice made in indexing compromises some aspect of the online catalog. At one extreme, there are the catalogs that don't allow direct searching at all. Every search is really a browse of a particular heading file, most commonly, authors, subjects, and titles. The technique is easy for first-time users and can work reasonably well for smaller collections but becomes extremely cumbersome for large collections and is always a relatively slow way to find a known item.

Slightly less extreme are the catalogs that lump everything into one massive word index. A patron keys as many words as seem appropriate, and the computer searches for items that have all of the words somewhere within the bibliographic entry. A sophisticated version of this technique will yield multiple result sets: those items that contain all of the words, and other lists of items matching some, but not all of the words. The proponents of this technique are vocal and, to some degree, convincing.

But an overall word index, particularly for a large collection and particularly if notes and other fields are included, has some interesting consequences for precision and recall. Some fairly common words become essentially useless as search terms, even though they could be used quite well for title-phrase searches. Keying one or a few words may result in an unmanageably large result set. Keying too many words may well eliminate desired materials. The choice here is to minimize the need to explain and understand indexes, at a significant but not overwhelming compromise in flexible access.

Tradeoffs and compromises continue in cases where multiple indexes are offered. Every additional index slightly increases the complexity of the user interface (there are more choices to be explained, and index names need to be longer), but additional indexes also increase searching power and flexibility. Does the catalog only include phrase indexes for titles and subjects—that is, retrieving only on the full access point or a leading portion? Phrase indexes ruin recall for people who don't remember the item properly and make subject searching more difficult. Does the catalog only include word indexes for these access points? That makes very short titles more difficult to retrieve, particularly in large collections. Does the catalog include both word and phrase indexes? That doubles the number of indexes, making them difficult to explain and remember. A typical online catalog for a large academic library probably has too few indexes and, simultaneously and paradoxically, too many indexes.

Flexible Interfaces vs. Maintainability

The beginning of this article mentions the choice between a flexible interface, one customized for each library or location, and a standard interface that can be maintained more easily. That choice represents a real problem that haunts most responsible vendors of online catalogs. Most of the catalogs have chosen flexible interfaces, and vendors have not had much choice in the matter. Library requests for proposals will tend to require such flexibility. The libraries may be paying a high price for their customized interfaces, as development time and money goes into trouble-shooting that is made much more difficult by the wide range of interfaces. Also, every customer would like to believe that a properly designed piece of software will be fully tested before it is shipped. But the number of permutations available within a typically flexible online catalog design is such that no vendor can afford to run an exhaustive set of tests on all possible cases; in this case, flexible design removes the possibility of a complete assurance of correctness.

The Cheat Sheet: Handout Documentation

"A good user interface doesn't need printed documentation." A good online catalog should offer a labeled key or soft-labeled function key that will call up enough information to get a patron going and will allow them to keep going. But the learn-while-you-use method fails two classes of patron: true computerphobes and those afraid that they will be "beeped at" or made to look silly will be much more comfortable if they can get some introduction to the computer before actually approaching the keyboard; and "power users," including those who really want to spend their time finding things, not learning the system, would rather spend five minutes with a well-prepared "cheat sheet" than fifteen minutes "playing around" at the terminal. Since there are never enough terminals, this time savings also saves hassles for other patrons.

There are at least three categories of written materials:

1. Complete guides to the system, unlikely to be publicly available in most OPACs and certainly not easy to hand out. A complete guide to a good OPAC would probably be book length.
2. Multipage brief guides, flip charts, etc. These should be available at or near the terminals, and should cover all functions of the system but require their own sets of compromises. A brief guide should have at least one example for each significant function, and should show some actual searches.
3. "Cheat sheets"—function and control summaries. These should be made cheaply enough that people can pick them up, and should usually be no longer than both sides of a single card or sheet of paper. They should not give complete information; instead, they should contain a list of commands and indexes or similar appropriate information. A cheat sheet should be simple enough so that a patron can refer to it while using the catalog without particularly thinking about the cheat sheet—ideally, help is just a glance away.

COMMENTS ABOUT CHOICES

Only a few of the hundreds of choices required to design an online catalog have been mentioned in this article. One way of summing up all the choices is to repeat that an online catalog cannot be all things to all people, and cannot possibly include every desirable feature. An online catalog should not simply be a collection of features.

Good designs reflect choices made within a consistent context. Every good online catalog presents a coherent model, an overall interface that establishes a clear pattern and follows that pattern. When a patron understands how part of the online catalog works, the patron can apply that knowledge to the rest of the system with good results. There are also compromises to make along that path, to be sure; patron expectations from early use will certainly not always mesh totally with the catalog design.

The best choices and the worst compromises both spring from an overriding model. The worst compromises come about when a model is designed without considering the range of alternatives and possible extensions to the catalog. For example, a model that relies entirely on single-digit numbered menus will become awkward when added indexes or added functions really require more than nine possible choices in certain situations. An inappropriate or limiting model will foreclose options for implementation and extension; the catalog will never become more than its original design, and that design will probably not serve very long or very well. In practice, catalogs based on inappropriate or overly limiting models are extended and enhanced, and the extensions violate the original model, leading to an incoherent, confusing user interface.

But the best choices also come about with deliberate models. Open-ended, flexible models guide choices and compromises but are much less likely to mandate particular choices and rule out additional features. Looking at the better models, one can see the ways that extended functions could be added without affecting the coherence of the current model and, to some extent, the shape of the extended functions can be predicted.

The User-Friendly Trap

One catch phrase for human/machine interface design is "user-friendly." But what is friendly for one user and one type of use may be hostile to another user and another type of use. Too many bad design decisions are defended on the grounds of user-friendliness.

Thanks to widespread use of personal computers, there is much more real-world experience with human/computer interfaces now than a decade ago. Many of the best software designs are changing in ways that, by traditional standards, might be considered less user-friendly. It is also clear, based on the range of successful products in that field, that different users have very different ideas of user-friendliness.

For example, word processing programs for PC-DOS/MS-DOS are probably the category with the largest total installed user base and certainly a category with no single dominant factor. Large-selling programs include heavily menu-driven interfaces with ever-present prompts; totally command-driven programs with very little on-screen prompting; menu-driven systems that show no information on the screen at all, leaving the user to find the right function keys to call up menus; and a number of other designs. The current bestseller, WordPerfect, trades a totally clear screen for a totally hidden set of wholly non-mnemonic menus. The most widely used (WordStar) and another powerful current design (Microsoft Word) both start out with a significant portion of the screen taken by command prompts; both also permit the user to eliminate the prompts with two or three keystrokes. Another widely used system,

PC-Write, leaves one very small prompt on the screen at all times; that prompt brings up function-key menus and dense help screens. Yet another, XyWrite, does almost everything through keyed commands, with little or no visible help. All five word processors work well, and there are large user communities who regard each of them as being user-friendly—and other users who regard them as being hostile and poorly designed.

It all depends on one's point of view. A pure menu-based online catalog may be friendly to a new user looking for a topic, but it will be less friendly for a patron looking for fast status information on a known title. That patron doesn't want to go through seven or eight screens to get at one piece of information when a single command could do as well. It will be even less friendly for an experienced researcher; the menus and prompts will clutter the screen and the inflexible methodology will slow searching and retrieval.

In other words, "user-friendly" is meaningless as a design criterion, at least by itself. Some choices are clearly user-hostile, but many choices can only be considered friendly or hostile in a particular context. Well-designed systems accommodate diverse patron needs; in the long run, that is the friendliest of all.

Give The Patrons What They Want

If user-friendly design is one trap, another is that catalogs should give the patrons what they want. Put another way, future research projects should investigate what features patrons like and dislike, what features they would like to see, and use the results as the basis for catalog designs. That is a good idea (one partially carried out in the Council on Library Resources online catalog study (Matthews et al., 1983)), but it is certainly not a total solution.

The problem with studying patron behavior and desires is that, by and large, people will do a better job of reacting to what they have than of considering what they could have. Further, people's ideas of what would be desirable don't always match up with the systems they find most useful. The best ideas for enhanced user interfaces are likely to come from people with solid backgrounds in libraries and experience with human/computer interfaces, who can come up with feasible ideas that patrons would be less likely to envision.

The value of feedback from patrons should not be minimized. Given a working system, a few of the more vocal patrons will let the library know what works and what doesn't work. Formal research projects may yield useful results. There were some useful results from the CLR study, although it suffered from three problems: first, many of the systems studied weren't really online catalogs; second, people dealing with the results over-interpreted them in some rather startling ways; and third, a series of relatively small samples resulted in a level of analysis and publication that was probably overkill. It would take a series of studies like the CLR study, done on larger scales (that is, with statistically significant samples for each catalog design) and over time to establish a body of reliable research on which to base catalog designs. Unfortunately, there appears to be no current likelihood of such studies being mounted.

Giving the patrons what they want is a good idea if patrons have the background and range of possibilities to know what they really want—i.e., what would serve them the best. A different rallying cry is "give the patrons tools they can use" which may underestimate the intelligence or learning ability of patrons. A surprising number of online catalogs treat library patrons like small children in ways that most small children would find objectionable. Many patrons have learned to use card catalogs effectively without formal instruction. It should be obvious that many patrons can think for themselves, and want tools they can use effectively and rapidly.

Personal Experience with User Feedback

The experience that the author had from designing an online retrieval interface was instructive, all the more so because the design was worked out in conjunction with representatives of the intended users who are not librarians and are not sophisticated computer users.

At the beginning of the design process, it was assumed that the system would use menus of some sort, since most users of this system would be new to this sort of thing. After some discussion, specifically noting that this particular database could not be searched at all well without heavy use of multiple-index Boolean logic, it was agreed to set out a command-language strategy and follow it with menu-driven or heavily prompted alternatives. The draft Standard Common Command Language was used as a basis for the command-driven version, first, because it is relatively similar to RLIN, and second, it has some possibility of becoming a standard used in a variety of retrieval systems. It also uses good syntax and a reasonably good set of commands.

A demonstration program showing the overall user interface (a three-part screen), the set of commands and normal syntax, online prompts and some other aspects, was sent to the representatives with a note that the menu alternative would be next. After working with the demo and talking it over, the representatives suggested not to provide a menu-driven alternative; with good online prompts, they felt that new users would find the command syntax perfectly reasonable to use, even for the first time. They also recognized, very rapidly, that the command syntax would permit much more rapid and much more fluid operation than a menu-driven alternative.

There are some disadvantages in a command-driven system. Menus are easy to implement and make keyboard handling quite simple. The keyboard handling for a command-driven system is not so easy and the search parsing routine is by far the most complex piece of code in the entire system, even though it currently omits some features that really should be included.

The interface included partial lists of actions displayed on each screen, together with a labeled function key to bring up a full set of actions. After using the system for a while, the representatives told the author to clean up the screen by eliminating the prompted actions; they felt that scholars would find the function key sufficient and would prefer a less cluttered screen.

Put another way, the way to design a system is *not* to ask the users what they want. If that had been done, the system would be a cumbersome mass of menus. Users like what they see now, but it wasn't what they would have asked for. User feedback is important, but user preferences can only be part of the design process.

CONCLUSION

The history of online catalogs is a history of compromises between what is ideal and what can be provided in the real world. Limitations based on economics may become less significant as computers become more powerful, but online catalogs will always involve difficult choices among various desirable features and attributes that conflict with one another.

A coherent model is at the base of every good online catalog, and a well-designed model makes choices clearer and can guide designers to the needed compromises. Without a coherent model, choices become haphazard and the catalog can never be fully effective. There is no single best model for an online catalog, but good models have some things in common.

Good models are extendable—new functions can be added without destroying the clarity of the model. That's not true of weak or badly defined models; such catalogs become collections of poorly related features as new functions are added. Good models are predictable. Once a patron has learned to use several of the functions in a catalog, they will approach other functions with a set of expectations; if those expectations are fulfilled, they will pick up the remaining functions very rapidly. Good models represent conscious intelligent compromises toward the ultimate aim of any online catalog—to provide the information that a patron needs in a way that suits the patron.

REFERENCES

Crawford, W. (1987). *Patron access: Issues for online catalogs.* Boston, MA: G. K. Hall.

Crawford, W.; Stovel, L.; & Bales, K. (1986). *Bibliographic displays in the online catalog.* Boston, MA: G. K. Hall.

Matthews, J. R.; Lawrence, G. S.; & Ferguson, D. K. (Eds.). (1983). *Using online catalogs: A nationwide survey.* New York: Neal-Schuman.

LESLIE EDMONDS

Youth Services Coordinator
St. Louis Public Library
St. Louis, Missouri

Starting Out Right: The Effectiveness of Online Catalogs in Providing Bibliographic Access to Youth

INTRODUCTION

For most librarians and information scientists, the ability of children to access library material or, indeed, information in general is an abstraction. It is seen as an issue of intellectual freedom, something "they" should teach in school or an ability to be taken for granted when helping the young patron in a library setting. For a parent, teacher, or youth services librarian, it may be a source of irritation that kids may not be particularly interested in using the card catalog. The library profession has not looked carefully at how children do use catalogs, nor has it been too honest in its appraisal of just how useable online catalogs are for children. This paper will present the findings of a research project that was designed to ascertain what knowledge children bring to using the catalog, what developmental skills children have that would help them use the catalog, and what success children have with known-item catalog searches. The methodology used in this study may also provide a model for examining the behavior of other user groups for whom interfaces are designed.

As W. David Penniman has suggested earlier in this proceedings, the work of accessing information is done in a social context. The culture of childhood in America is complex, with a legal system that treats youth as a special case, including what access youth may have to information. In the U. S. education system, economic factors have created an interesting gap in opportunities for children. The increasing number of poor and homeless children have little access to information because of meager environments at home and in school. The social context of children seeking information should not be romanticized. Right now the best weapon against AIDS, drug abuse, and teen parenthood is information: getting kids to know and face the facts.

To begin to look at how institutions, specifically public libraries, might begin to meet both traditional and survival information needs of children, through appropriate catalog access, it might help to sort out some myths from reality.

Myth #1: All children like computers, so all children will like and use online catalogs.

While children may be more used to computers than adults, it appears that computers are still more interesting to and more manageable by males than females. There is not much data for gender difference with elementary school-aged children, but the gap becomes noticeable in the middle grades and widens in high school (Sanders & Stone, 1986, p. 5). Computers are more likely to be available to affluent children than to poor ones. It may also be stretching the truth to assume that children who enjoy computer games or even computer-assisted instruction will necessarily find an online catalog desirable simply because it is online. That is like saying reading a card catalog is as much fun as reading fiction. Even most librarians don't find this to be true.

Myth #2: Children naturally want to be good information managers.

While it is to be hoped that children want good information, it is unlikely that they enjoy the challenge of finding the information they want. The artistry of online search capabilities is beyond most children. A full MARC record is unlikely to be interesting to a young catalog user who lacks the sophistication to understand much of what is included and whose goal is very concrete. A child is more likely to want the good book that a friend was reading or to be told the capital of Wyoming for the report due tomorrow than to care much about knowing the right subject heading or being able to power search using natural language. The librarian's goals for library catalogs are often not the same as the child's goals.

Myth #3: Teaching alphabetization will allow children to be successful online catalog users.

It is common to teach children how to alphabetize and to use a few simple filing conventions as part of a library skills curriculum (Bell

& Wieckert, 1979; Welken, 1967). It is not clear that there is a direct relationship between these activities and independent use of an online catalog. It is likely that children need to understand and practice when to apply the rules, but learning alphabetization is not enough to insure both a successful and an efficient catalog search. This may also be a case of too little, too early. Alphabetization skills are not enough for touch screen designs and not really needed for most keyboard systems (spelling is probably more important for either system) and children seem to have trouble with the conceptual relationships involved in catalog design and use until they become comfortable with logical thinking, at about age twelve.

To explore some realities about bibliographic access for children, research was conducted at the Downers Grove (IL) Public Library (Edmonds et al., 1989). This project was funded by the Carroll Preston Baber Award which is given by the American Library Association to support innovative research on aspects of library technology. As there was no other research on children and online catalogs that could be located, nor recent research on children's use of the card catalog, it seemed important to expand the project to include children's use of both card and online catalogs, so that a comparison of the two could be made.

The need for evaluation of online catalogs goes beyond acceptance or comfort in the use of the computer. Accepting computer technology and using hardware is a first step to using online catalogs, but the more important question is whether children can actually use existing online programs, i.e., the software, to find needed materials. Of particular concern are the developmental skills required by the interactive software. Can children understand information presented on the catalog screen? Are they generally capable of understanding search methods to be used? Does the online catalog offer any advantages or disadvantages when compared to the traditional card catalog? Answers to these questions will help software designers and librarians define the parameters of effective use of online catalogs by children, and plan modifications in the online catalog, in bibliographic instruction, and in reference/reader's advisory services.

RESEARCH PLAN

This study evaluates children's use of an online catalog to gain bibliographic access to materials at the Downers Grove Public Library (DGPL). The study also evaluates the children's ability to use the DGPL card catalog. Although most libraries close their card catalogs shortly after opening the online catalog, the DGPL chose to maintain its card catalog pending resolution of some specific problems associated with the online database. DGPL touch terminals were used to access a CLSI online catalog (OPAC terminal) (Rice, 1988, p. 14). A specific set of tests was developed to determine if the children had the necessary skills to effectively locate and interpret bibliographic information as presented by both forms of the catalog.

A sample of children in fourth grade (nine and ten year olds), sixth grade (eleven and twelve year olds) and eighth grade (thirteen and fourteen year olds) was tested to see if their skill development would allow them to follow the online catalog protocols and interpret information presented on the screens in order to identify materials in the library's collection. In addition to observing student use of the library's online and card catalogs, general skill level was measured by a written test (see Appendix A). The children were also asked about their preferences about online versus card catalog use. The skills test included items on alphabetizing individual words, names and phrases, and applying simple filing rules for ordering titles, authors and subjects. During the observation/interview, randomly selected students demonstrated actual skills in manipulating the card or online catalog. The research observers noted the child's ability to find call numbers for known items held by the DGPL and the child's efficiency in finding bibliographic data (see Appendix C).

Students were given a card with a title, author, or subject heading printed on it and were instructed to find in the catalog the specific title, any book by the listed author, or any book on the given subject. Each student had the opportunity to do a title, author, and subject search, though the student could decline to do more than one search or the observer could excuse the student who was not making progress in a search. For searches in either catalog, students were asked to find the call number once the search item entry was located. If the student said the call number aloud or pointed to it, the search was considered a success.

Sample

Each grade was represented by over fifty students and the population was roughly divided into thirds for the fourth, sixth, and eighth grades. The sample was also fairly evenly divided by gender. The students reported being regular users of the DGPL with about 68 percent using the library once or more a month and about 10 percent never having used it. The older students were fairly familiar with the OPAC terminals and very familiar with the card catalog. Only about a third of the fourth graders had used the OPAC previously. More males had used the OPAC, but males and females had used the card catalog about equally. The students in the study had received instruction in library skills as part of a district-wide school library instruction program in which fourth graders are introduced to the card catalog and sixth graders are introduced to simple filing rules.

Preference Survey

Children perceived that the card catalog was easier to use, perhaps because they were more familiar with it. Ease of use was the predominant reason for stating preferences. The majority of children in all grades and of both genders preferred using the card catalog over the OPAC terminal (68 percent to 16 percent). The remaining 16 percent had no preference. Lack of familiarity with the OPAC was often the reason given for using the card catalog, implying that the card catalog was preferred by default. While most students were able to express an opinion about catalog preference, there seemed to be no strong allegiance to either format.

Skills Test

In general, sixth and eighth graders demonstrated moderate knowledge of alphabetizing and filing rules, but more than fifty percent of the fourth graders were unskilled (Table 1). There was very little difference in performance by gender. All students did well on simple alphabetizing, but had difficulty with questions that required knowledge of specific filing rules or multiple-word phrase alphabetizing. On the basis of the skills test, it would be predicted that students would have difficulty using either the card catalog or the OPAC. It could also be expected that fourth graders would be the least successful, while the eight graders would have the best success rate.

TABLE 1
TOTAL TEST SCORE BY GRADE

	Grade			
	4	*6*	*8*	*Total*
Skilled (95-115)	5 6.17%	6 9.38%	13 20.97%	24 11.59%
Moderately Skilled (75-95)	35 43.21%	47 73.44%	48 77.42%	130 62.80%
Unskilled	41 50.62%	11 17.19%	1 1.61%	53 25.60%
Total	81 100.00%	64 100.00%	62 100.00%	207 100.00%

Observations

In the card catalog observations, fourth graders were much less skilled in locating the call number than the older students (Table 2). Using the OPAC, students were far less successful (Table 3). There were 65 percent successful searches with the card catalog as compared to almost 18 percent success with the OPAC. No fourth grader was successful with the OPAC. As well as gathering data on success rate, data were collected on the effectiveness of both the card and online searches. Based on the number of touches online or the use of the card guides in the card catalog, sixth and eighth graders were moderately efficient (i.e., made fewer mistakes and had to start over less often) and fourth graders were inefficient. Students were more efficient when using the card catalog than when using the OPAC. Females and males had the same patterns of performance during the observations.

TABLE 2
CALL NUMBER IDENTIFICATION-CARD CATALOG BY GRADE

	Grade			
	4	*6*	*8*	*Total*
Correct call number	27 33.33%	83 70.94%	103 79.84%	213 65.14%
No call number given	54 66.67%	34 29.06%	26 20.16%	114 34.86%
Total	81 100.00%	117 100.00%	129 100.00%	327 100.00%

TABLE 3
CALL NUMBER IDENTIFICATION-ONLINE CATALOG BY GRADE

	Grade			
	4	*6*	*8*	*Total*
Correct		20 25.64%	17 27.42%	37 17.96%
Wrong	66 100.00%	58 74.36%	45 72.58%	169 82.04%
Number of attempts	66 100.00%	78 100.00%	62 100.00%	206 100.00%

RESEARCH FINDINGS

The children in this study did not have knowledge of many rules, nor did they seem to be able to understand useful concepts (e.g., between, doing things in a standard order or sequence, being precise) well. Näive users of any age will need to increase knowledge, practice sequencing, and concentrate on accurate task performance. The older students were better able to correct errors and were more knowledgeable than the younger ones, though few students demonstrated any sophistication in catalog use. It was found that, because use of the catalog can be complex, children who are not yet developmentally capable of mastering necessary logic for the catalog may require simpler library catalogs.

The test of skills, as well as the selection of sample search items, simplified the task of a student finding necessary information. Students were not asked to accomplish complex searches. Given that students did not demonstrate mastery of simple and moderately difficult rules and searches, this choice seems to have been a reasonable one. But one needs to be mindful that if students enter the library to find materials, their "real" searches will not necessarily be as straightforward as the ones presented to them in this project. Logically, one would expect students to have a lower success rate when they are performing more complicated searches. For example, students were not asked to demonstrate knowledge about how numbers are treated in a title or about catalog conventions for eras in U. S. history. Yet students might need information on the Civil War or need to locate *1984* by George Orwell. It is appropriate to view skill attained by subjects in this project as the upper limit of performance. It is unlikely that these students would perform better on a wider variety of searches that represent their real information needs.

While the findings of this research have implications for both bibliographic instruction and reference service, the findings about interface design are more closely related to this conference. Instructional plans and changes in the reference interview can be adapted equally well for both the card catalog and the OPAC. The OPAC, however, is much more receptive to design changes than the card catalog. Although subjects were more skilled at using the card catalogs than the OPAC, it is not reasonable to conclude that the card catalog should automatically be the catalog of choice for children. As design changes are made, OPAC will become the appropriate choice.

There are three areas where interface design could be changed to improve bibliographic access to children. Initially, the sequences or steps presented should be reduced and simplified. Children have as many as twelve screens to manipulate and five different screen designs to interpret. Since the card catalog offers four basic steps (file selection, drawer selection, card selection and card interpretation), the OPAC should be designed to parallel more closely this simpler model.

Secondly, the individual screen designs need improvement. Since children have some trouble searching alphabetically, it would be helpful to reduce the number of choices offered at one time. The OPAC screens offer eleven terms to read and thus eleven choices to make. This can be overwhelming. A series of natural language questions such as "Does your word come alphabetically between hat and hot?" would be easier for children than presentation of word lists. Children had trouble understanding the entry screen. It contained too much information and too high a percentage of print on the screen. The lines should be placed farther apart with larger print and less information. It may be possible to present the patron with a simple bibliographic entry, such as on a bibliography, and allow access to the shelf list on a second screen. Unsophisticated patrons rarely need a full citation. The other screen design needing improvement is the help screen. Help messages should be tailored to each kind of error that can be made, and assistance targeted to the particular error only. Generic help screens ask the patron to read through extra information to determine what steps are necessary to proceed.

The third area of concern in interface design is that instruction and/or coaching should be provided online. Front end instruction can be designed to lead patrons through a variety of search possibilities. If patrons are knowledgeable users, this instruction can be bypassed, or, if users find they need help in the middle of a search, they can move to a diagnostic mode ("Are you sure of the spelling of your search term?", "What other, related terms could you use?") or a coaching lesson. The great strength of computers is that they may be used in different ways by each user, but this capability is not evident in this OPAC system.

The findings of the Baber Research Project present librarians with an opportunity to consider children's accessibility to a library's bibliographic records and the difficulties they encounter. These initial efforts to isolate and quantify the steps taken by users of library catalogs need to be replicated in other settings. For software designers and online catalog vendors, the information gained may provide a framework for future improvements. Based on this research, there can be a heightened sensitivity and awareness for the multiple problems experienced by children attempting to locate materials using library catalogs.

CONCLUSION

While this research found no real gender difference in performance on the OPAC, or any other measure, there seems to be some evidence that not all children approach the OPAC in the same way. Grade level and developmental level seem to be predictors of how successful children will be in using library catalogs. The observations showed how children dealt with frustration. Some gave up, some moved into an almost frenzied activity mode and did not want to stop, and others stepped back and tried to figure out the rules or concepts that would lead to success. Individual personality seems to have some effect on how children used the OPAC. The children, generally, were very excited about being in an experiment. (They seemed to expect a laboratory and Dr. Frankenstein!) They were cooperative, but expressed frustration with the OPAC and did not seem swayed for or against the catalog because it was online. Feelings about computers in general do not seem to affect use of the OPAC.

It also seems that children are different from adults in being able to understand directions, in doing things in order, and in being precise about what they are doing. One OPAC direction is, "Choose the word that comes alphabetically before the term you are looking for." Even the youngest students were able to do this, but if the student was looking for the word "octopus" he or she might start with "kangaroo," because "k" comes before "o." Since the database is large, this lack of precision leads to lack of success. Many children tried to use the help function, but couldn't choose the correct help message to solve their particular problem. It would be interesting to observe adults to see how they would differ from our young sample.

Though these children had been instructed on how to use the catalog and had some ideas about filing rules from school, they were not adept at applying them at the public library. They used a card catalog at school but had no real idea that some of the same rules might apply on the OPAC. Since the skills test used was made up of items from standard library skills curricula and children had moderate mastery of those items, it would suggest that what is taught and what is needed differ from one another.

Though this study is very preliminary, a few suggestions may be in order as to what might be done to make children more successful OPAC users. First, of course, there needs to be further study using other online catalogs, both replication of the outlined method and the development of other ways to evaluate user competence. Vendors need to field test online catalogs with children to see what help is needed and to adapt screen design to this user group's level of understanding. Systems need to use the strength of computers to build in flexibility, so each user has choices about how much help he or she needs, how fast he or she can proceed, and how full an entry is needed. Librarians need to be realistic about how much help children will need. "Go look it up" has long been discouraged as part of the reference interview, not so much because it is rude, as because it is an impossible task for many young patrons. Lastly, librarians and vendors need to address the patron's need for useful instruction. For children this instruction needs to be multifaceted, including both concepts and details. It needs to be a part of the reference interview and imbedded in the online catalog.

While much is still unknown about how people use catalogs, a lot is known. Librarians need to take steps to design catalogs based on user needs rather than technical capability. Patrons are more than ready for easier-to-use catalogs.

APPENDIX A

ALPHABETIZING AND PREFERENCE SURVEY
BABER RESEARCH PROJECT
DOWNERS GROVE PUBLIC LIBRARY

NAME ______________________________

SCHOOL ______________________________

GRADE ______________________________

BOY __________ GIRL __________

DATE ______________________

1. In the last year, how often have you visited the Downers Grove Public Library?

    _____ More than once a month

    _____ Once a month

    _____ Less than once a month

    _____ Never

2. Have you ever used the PAC terminals (online catalog) to locate materials in the Downers Grove Library?

    _____ Yes _____ No

3. Have you ever used the card catalog to locate materials in the Downers Grove Public Library?

    _____ Yes _____ No

4. Which catalog do you prefer to use?

    _____ PAC terminal (online catalog)

    _____ Card catalog

    _____ No preference

    Why? ______________________________

    ____

    ____

1. Write the letter that comes *before* each of these letters in the alphabet.
   Example: _____ z

_____ c   _____ m

_____ q   _____ i

_____ o   _____ v

_____ j   _____ e

_____ r   _____ x

2. Write the letter that comes *after* each of these letters in the alphabet.
   Example: a _____

f _____   u _____

w _____   k _____

p _____   d _____

h _____   n _____

s _____   l _____

3. Number these words in alphabetical order.

Example: _____ cat

_____ dog

_____ apple

_____ used

_____ wear

_____ vine

_____ tear

_____ quiet

_____ open

_____ pour

_____ robin

4. Number these words in alphabetical order.

_____ dove

_____ dwelling

_____ dear

_____ dyed

_____ dry

_____ during

_____ date

_____ diet

5. Number these words in alphabetical order.

_____ poster

_____ porcupine

_____ power

_____ poultry

_____ possess

_____ porcelain

_____ powder

_____ positive

6. Authors and titles.

Does *Little Pear* come before or after "Little, Jean" in our card catalog?

_____ before

_____ after

Does *The Bell Witch* come before or after "Bellairs, John" in our card catalog?

_____ before

_____ after

Does *Wildlife on the Watch* come before or after "Wilde, Oscar" in our card catalog?

_____ before

_____ after

7. Pretend these are labels on the card catalog. Write down the drawer number where you would look for each author or title.

1. A-B	4. Dd-Do	7. Hep-Jan	10. M-Mac
2. C-Ch	5. Dp-Goh	8. Jao-La	11. Mb-Mi
3. Ci-Da	6. Goi-Heo	9. Le-Lz	12. Mj-Mz

1. *Moominland Winter* _____
2. *Is That You Miss Blue?* _____
3. *He's My Brother* _____
4. *Candles, Cakes, & Donkey Tails* _____
5. *Einstein Anderson, Science Sleuth* _____
6. McGowan, Tom _____
7. Fradin, Dennis _____
8. Hest, Amy _____
9. Dewey, Ariane _____
10. *Hepzibah* _____
11. *Mr. Popper's Penguins* _____
12. *Dr. Doolittle* _____
13. Goffstein, Brooke _____
14. *Gravity is a Mystery* _____

8. Number these titles in alphabetical order as they appear in our card catalog.

_____ Saints for All Seasons

_____ St. Patrick's Day

_____ The Saints

_____ Saint Jerome

9. Number these titles in alpabetical order as they appear in our card catalog.

    _____ Ah-Choo

    _____ The ABC's

    _____ A Horse Called September

    _____ A is for Angry

    _____ The Horse and His Boy

10. Number these titles in alphabetical order as they appear in our card catalog.

    _____ Is There Life on a Plastic Planet?

    _____ Dr. Doom: Superstar

    _____ Deathwatch

    _____ Look-it-up Book of Stars and Planets

    _____ Look! Look!

11. Number these titles in alphabetical order as they appear in our card catalog.

    _____ Monster is Coming

    _____ Monster in the Third Dresser Drawer

    _____ Monster Manners

    _____ Monster Mania

    _____ Monsters, Mysteries, and UFOs

    _____ Monster Trucks and Other Giant Machines on Wheels

    _____ Monster Birthday Party

    _____ Monster & the Tailor

    _____ Monsters from the Movies

    _____ Mountains Around the World

    _____ Monster's Nose Was Cold

12. Alphabetizing authors.

Does "Fleischman, P.R." come before or after "Fleischman, Charles" in our card catalog? _____ before

_____ after

Does "Flesch, Y." come before or after "Fleischman, Charles" in our card catalog? _____ before

_____ after

Does "Fletcher, Jessica" come before or after "Fletcher, A. Sarah" in our card catalog? _____ before

_____ after

13. Number these names in alphabetical order as they appear in our card catalog.

_____ Foster, Stephanie

_____ Fortune, J. J.

_____ Ford, Ford Madox

_____ Foster, Sally

14. Number these subjects in alphabetical order as they appear in our card catalog.

_____ Carpentry

_____ Carnivals

_____ Castles

_____ Cartoons

15. Number these authors and titles in alphabetical order as they appear in our card catalog.

_____ *Frankenstein*

_____ Frankel, Max

_____ *Frank and Mary*

_____ Fradin, Dennis

16. Number these authors and titles in alphabetical order as they appear in our card catalog.

    _____ *Amos Fortune, Freeman*

    _____ Freeman, Don

    _____ *Free to Be You and Me*

    _____ Fortune, Amos

17. Number these subjects in alphabetical order as they appear in our card catalog.

    _____ Foods—Manufacturing

    _____ Friends, Society of

    _____ Food—Social Customs

    _____ Friendship—Ethics

APPENDIX B

SEARCH ITEMS

TITLE

Who Will Be My Friend
The Poison Factory
Octopus Pie
The Best Bad Thing
The Curse of the Blue Figurine

AUTHOR

Fairman, Paul W.
Harder, Eleanor
Geibel, James
Chaconas, Doris J.

SUBJECT

Animals, Training of
Fire Stations
Seals (Animals)
Payton, Walter, 1954
Insects — Poetry

APPENDIX C

ONLINE CATALOG OBSERVATION RECORD

Name ____

M _____ F _____

Grade _____

Search Card #1 _____ #2 _____ #3 _____

Date ___/___/86

I. Identify the elements of the online catalog entry:

AUTHOR _____ Immediate _____ With Difficulty _____ Unable

TITLE _____ Immediate _____ With Difficulty _____ Unable

CALL NUMBER FOR DGPL COPY _____ Immediate

_____ With Difficulty

_____ Unable

IS THIS A BOOK DGPL OWNS? HOW DO YOU KNOW?

_____ Immediate

_____ With Difficulty

_____ Unable

II. Search Card #1 _____

1) DIFFERENTIATES AUTHOR-TITLE-SUBJECT

_____ Yes

_____ No

_____ Correct 2nd Time

_____ Unable

2) LOCATES CORRECT ENTRY _____ Yes

_____ No

3) CALL # GIVEN ______________________________

UNABLE TO LOCATE CALL # ____

GIVES CLASS # ____

4) NUMBER OF TOUCHES ____

5) NUMBER OF TIMES STARTS SEARCH OVER AT BEGINNING ____

6) TIME OF COMPLETE SEARCH ________________

SEARCH INCOMPLETE ____

Comments:

APPENDIX D

CARD CATALOG OBSERVATION RECORD

Name ______________________________

M ____ F ____

Grade ____

Search Card #1 ____ #2 ____ #3 ____

Date ___/___/86

I. Identify the elements of the online catalog entry:

AUTHOR ___ Immediate ___ With Difficulty ___ Unable

TITLE ___ Immediate ___ With Difficulty ___ Unable

CALL NUMBER ___ Immediate ___ With Difficulty ___ Unable

II. Search Card #1 __________

1) DIFFERENTIATES AUTHOR-TITLE & SUBJECT

____ Immediate

____ Begins on wrong side:corrects

____ Unable

2) CHOOSES CORRECT DRAWER

____ Immediate

____ Near miss on locating correct drawer

____ Number of incorrect attempts

____ Random attempts to locate drawer

____ Unable to locate

3) CHOOSES CORRECT CARD

____ Locates correct card

____ Uses guide, efficient search

____ Appears to use guides occasionally

_____ Card by card

_____ Flips randomly through drawer

_____ Unable

4) CALL NUMBER GIVEN ______________________________

Comments:

REFERENCES

Bell, I. W., & Wieckert, J. E. (1979). *Basic media skills through games.* Littleton, CO: Libraries Unlimited.

Edmonds, L.; Moore, P.; & Balcom, K. M. (1989). *An investigation of the effectiveness of an online catalog in providing bibliographic access to children in a public library setting* (Report funded by the 1986 Carroll Preston Baber Research Award, March 1989). Urbana-Champaign: University of Illinois, Graduate School of Library and Information Science.

Rice, J. (1988). The dream of the memex. *American Libraries, 19*(1), 14.

Sanders, J. S., & Stone, A. (1986). *The neuter computer: Computers for girls and boys.* New York: Neal-Schuman.

Welken, M. L. (1967). *A guidebook for teaching library skills: Book 4.* Minneapolis: T. S. Denison.

ROSE MARIE WOODSMALL

Program Analyst
Office of Planning and Evaluation

ELLIOT R. SIEGEL

Assistant Director for Planning and Evaluation
National Library of Medicine
Bethesda, Maryland

Reconciling Design Philosophy and User Expectations

INTRODUCTION

This paper discusses the design, development, and evaluation of GRATEFUL MED, the National Library of Medicine's (NLM) front end software for microcomputers that was developed to assist physicians and other health professionals to search NLM's MEDLINE database. A search is constructed by filling out a form screen with information on the desired author, title, and/or subject(s); the search can be limited to English language, review articles, or a particular journal. No knowledge of Boolean connectors or the Library's Medical Subject Headings (MeSH) vocabulary is assumed. The search is constructed and the results reviewed on the user's microcomputer; that is, while not connected to the NLM mainframe.

By March, 1988, 9,500 copies of GRATEFUL MED had been sold. The owners comprised 38 percent of the MEDLINE searchers (by codes in use), accounting for 30 percent of NLM's mainframe computer usage.

Throughout the relatively short life of GRATEFUL MED, its designers have worked toward the principle of responding to the users' expectations while adhering to the original design philosophy—not always an easy task. This chronological look at the development of GRATEFUL MED shares some of the practical "lessons" learned along the way that may assist others interested in the design and evaluation of the computer/human interface. These lessons are presented from the personal perspective of the first author, not NLM.

"Design philosophy," in this context, refers to software functionalities rather than the ADP definition of what takes place between analysis and implementation. Reconciling the design philosophy and user expectations thus might be better phrased as "adjusting what you have decided to provide with what functions users (grow to) expect." It sounds rather recalcitrant when phrased this way; however, if the design philosophy is abandoned as a result of responding to the often diverse user recommendations, the software develops in a hodgepodge way, resulting in a deviation from the original goals. The key is "reconciling"—answering the user's need within the framework of the design philosophy.

THE BEGINNING

The first author's daily notes for December 14, 1984 say: "Upbeat meeting with John Anderson [Director of Information Systems] and Lois Ann Colaianni [Associate Director of Library Operations], and others. Dr. L. wants a concrete proposal for end user services."

Dr. L. refers to Donald A. B. Lindberg, M.D., Director at the National Library of Medicine. This was one of his first directives; he had been sworn in only two months before. The author's specific assignment from this meeting was to survey existing end user search systems, including mainframe resident, microcomputer front ends, and front ends co-resident on a mainframe—both operational and prototypes—and prepare a report on the state-of-the-art that included recommendations for directions NLM might take in end user searching. The report was to identify existing commercial systems for possible endorsement or purchase by NLM.

Dr. Lindberg emphasized that the NLM system goal was to develop a searching mechanism for the end user that would obviate the need for the user to have experience with the three stumbling blocks to searching: logging on, using the ELHILL command language, and using the NLM controlled vocabulary, MeSH. In addition, it would be easy to use, inexpensive, and allow for a growing searching sophistication if the user so desired. He expressed a preference for a system that would require minimal documentation, perhaps a few pages, with extensive online help. Plans were for the evolutionary development of successive versions of the software, each "smarter" than the preceding one.

By February 2, 1985, the author's office walls were plastered with huge charts describing twenty-four operational and prototype systems, each described by a number of criteria, such as mode (command or menu), search logic, and key system features and capabilities. One prototype, called MICROSEARCH, was especially interesting, for it was designed to interface with software that had been derived from the NLM mainframe search software, ELHILL, which is used to search the NLM family of databases. Importantly, it appeared to satisfy all of the desirable characteristics identified by Dr. Lindberg—and then some.

All of the intellectual processes were done while the user was sitting at his microcomputer but not connected to the host mainframe computer. This feature was appealing because it freed the user from the sense of urgency that often occurs when the "clock is ticking" and charges are mounting. Automatic logon, built-in Boolean connectors, and an algorithm to recommend subject headings based on the user's judgement of the retrieval were other features of import. The design also included the concept of extensive online help. At the conclusion of the study in mid-February 1985, it was recommended and agreed that NLM would undertake with Online Information International the adaptation of the MICROSEARCH prototype to the NLM environment, a not-especially overwhelming task given its common NLM/ELHILL origin.

ACCEPTANCE TESTING OF SELECTED PROTOTYPE

The Office of Planning & Evaluation (OPE) was enlisted with providing a process to gather end user input for the evolving set of requirements for a MEDLINE front end based on the prototype. But before that, the NLM team had to do in-house acceptance testing of the MICROSEARCH prototype as it took on the look and feel of what was to become GRATEFUL MED. A highly interactive team approach between NLM and the prototype developer was thus begun at this early stage and continues to this day. This has been a major factor in the rapidity of the development. Later in the process, Online Information International was awarded a competitive contract to continue development, and is referred to from now on as the contractor.

Acceptance testing of the prototype was done by a small in-house group from MEDLARS Management Section (MMS) and OPE over an eight-week period. The group went through twelve versions in this short time, and major changes were made—cosmetic, functional, and philosophical. Written reports were compiled from testers' comments, and the contracter responded with feedback and fixes as rapidly as problems were found. Proposed enhancements were discussed by telephone and electronic mail. Thirty-two problems were identified and corrected; nineteen enhancements were considered, many of which were implemented and found to be useful.

Adherence to the design philosophy was a struggle throughout, even at this early stage when feedback was limited exclusively to in-house staff. For example, the team strongly considered allowing the user to limit a search to HUMAN, a frequently used search tactic in clinical searches (to eliminate animal research). This can be done by combining HUMAN, a legitimate MeSH term, with a search by a Boolean AND. It can also be done by combining AND NOT ANIMAL with a search. The front end could be constructed to do the search either way. However, experimentation showed that the more straightforward AND HUMAN increased the cost of a search sharply due to the ANDing of such a highly posted term. Searching with AND NOT ANIMAL, on the other hand, lost from 0 percent to 25 percent of the retrieval in a group of test searches—those citations indexed to both HUMAN and ANIMAL. A totally unexpected finding was that otherwise relevant citations indexed to *neither* HUMAN nor ANIMAL included a large number of articles (33 percent of the total) classified as "review articles." Since the software was intended for end users, many of whom would be doing searches to retrieve "a few good articles," it was felt that this potential elimination of review articles was likely to prove harmful. It was also felt that searches of a clinical nature frequently are limited to HUMAN by the very nature of the other terms in the search, making it unnecessary to add the term. An added factor was the significant workload that would be added to the mainframe if large numbers in the health professions adopted the front end for searching and elected to routinely (and simultaneously) limit their searches to HUMAN. Ultimately (this whole process spanned perhaps two days), it was decided to provide the user with an easy option to limit the search to REVIEW rather than HUMAN.

This rather lengthy example is given to illustrate one of the most important things learned:

> *Lession #1: Test every idea for enhancement thoroughly, using a group of searches if necessary, to measure the effects of the change—especially the unexpected effects.*

Each enhancement or change was thoroughly considered against the design philosophy before implementation in the evolving front end. This was to become the pattern throughout development of the ensuing versions.

Dr. Lindberg and other senior staff were shown the prototype several times, both routinely and when the group was dealing with difficult decisions. In addition, the prototype was demonstrated to several members of the NLM Long-Range Planning Panels (both health professionals and librarians), the NLM Board of Regents, and selected visitors of NLM senior staff. Some of these individuals were recruited as on-the-spot testers, and their reactions were invaluable. The acceptance testers had already grown too familiar with the prototype and were missing some obvious places where there was room for improvement. This is a natural tendency as patterns of use develop, which leads to:

Lesson #2: Add new testers periodically at all phases of development to avoid the problem of "pattern testing."

A recommendation for acceptance of the prototype was made in early October 1985. The front end was tentatively named "MEDSEARCH," and all on-screen and written references as well as the draft documentation used this name.

PROTOTYPE TO VERSION 1

Formal Beta Testing

A MEDSEARCH Beta Test Group was appointed by John Anderson in late September as acceptance testing was coming to a close. Besides the authors, Becky Lyon-Hartmann, Regional Medical Library Coordinator, and Edward Sciullo, from the Office of Computer & Communications Systems, rounded out the team that was to develop and carry out a Beta test plan.

More explicitly articulated design principles for both the software and the Beta test were needed to provide to the Beta testers, and ultimately, to the prospective users. They were produced as follows:

"The goal of MEDSEARCH is to provide a microcomputer software tool for end users and librarians that will:

— enable MEDLINE and CATLINE searches to be done with little or no knowledge of the presently required login procedures and search mechanics, or of the controlled vocabulary (MeSH);

— produce a reasonable number of citations that are responsive to the user's information needs and offer guidance in a more comprehensive search, if so desired;

— upload a search and download the results to minimize online time and attendant charges; and

— encourage, but not require, the user to learn more sophisticated searching techniques."

"The goals of the test were, in order: to pre-test the current prototype software to discover serious flaws or technical deficiencies prior to widespread distribution to the biomedical community; and, to obtain suggestions for future improvements and enhancements."

The Beta Test Group elected to use a phased approach to the Beta test. Phase I would consist of in-depth structured interviews and observations of actual search sessions with a cross-section of end users and librarians at each of three local test sites. In Phase II, the software would be distributed to up to fifteen sites for wider, but less tightly monitored, testing. The types of information to be obtained, in whole or in part, at each of the Beta test sites included:

— who the users are (professional role/specialty; computer experience);

— who the non-users are, and why;

— location of system (library reading room, hospital, laboratory, home, office, etc.);

— types of searches performed;

— user satisfaction with information retrieved;

— user satisfaction with the process;

— use of follow-on search capability;

— use of advanced front end and native search capability;

— problems, if any, in establishing log in;

— inadequate, incorrect, or inefficient documentation; and

— problems, if any, in installing software.

For the sake of consistency, one person was appointed to conduct the Phase I test using a packet prepared for collecting demographic information about the user, detailed information on the search itself, observations of the user/system interaction, information on user satisfaction, and suggestions and comments from the user. Structured questions were developed for the demographic and user satisfaction information; forms were outlined for the remaining open-ended information to be recorded by the observer. A two-minute "script" was prepared in order to deliver the same introduction to each test subject.

The NLM Reading Room was used as the initial site for Phase I Beta testing in order to pre-test the questionnaire and procedures. This goal was met, and a few minor changes were made to the questionnaire; however, the user population volunteering for the test at this site was composed of a disproportionate number of students in non-health fields and "private citizens," neither of which was the target health care professional population for which the system was conceived and designed. The reaction of the intended user population was anxiously awaited, and led to the rather obvious, perhaps,

Lesson #3: Test with the intended user population for the most useful feedback.

The two remaining Phase I sites, the University of Maryland Health Sciences Library and Fairfax (Virginia) Hospital Medical Library, co-ordinated by Frieda Weise and Alice Sheridan, respectively, provided useful feedback that confirmed that it was appropriate to proceed with Phase II. Thirty-five subjects at these two sites performed searches of their choosing without significant problems. The Beta test questionnaire that was to be used at the Phase II remote sites was also pre-tested with these users. There were a number of suggestions for enhancement, all duly recorded. It had been speculated that the information needs of the end users at the research site (Maryland) and the clinical site (Fairfax) would be quite different, and this was borne out. The clinical subjects were less interested in extensive retrieval (a "few good citations" sufficed), more time conscious, and keen on the idea of searching at home in the evening after seeing patients.

More feedback was wanted before deciding what substantive changes to make to the software, but there was enough observed behavior to make six small changes. One change did not seem particularly significant at the time—the addition of a Help screen for a search that retrieved nothing. This problem of zero retrieval became the focus of much more attention later on, thus:

Lesson #4: Pay close attention to every problem the user has; in the early stages, it is difficult to recognize what is really important.

Examples were changed and added in the *User's Guide,* which was truly an example of minimalism. Some of the expanded examples were searches done by Phase I testers; it added a note of realism to know that someone actually wanted information on these subjects.

Phase I testing was finished on November 25, and Phase II testing began. The seven Regional Medical Libraries (RMLs) had been asked and had agreed to participate: New York Academy of Medicine; University of Maryland Health Sciences Library; University of Illinois at Chicago, Library of the Health Sciences; University of Nebraska Medical Center, McGoogan Library of Medicine; University of Texas Health Science Center at Dallas Library; University of Washington, Health Sciences Library; and UCLA Center for Health Sciences, Louise Darling Biomedical Library. In addition, the following institutional sites had volunteered to participate in Phase II: McMaster University, Department of Clinical Epidemiology and Biostatistics, and Health Sciences Library; University of Missouri, Information Science Group, School of Medicine; University of Pennsylvania, Biomedical Library; and University of California, San Diego, Biomedical Library. In addition, several individual end users had volunteered. Some of the latter group would be sharing the software with others at their sites, including one clinic setting with sixteen staff and forty residents. Each site was contacted by telephone for a discussion of the goals, test methodologies, and timeframe. By November 27, the Beta test group was formed and the test was scheduled for mid-December through January, 1986.

At this point, the official name for the software was selected by Dr. Lindberg from a list of twenty-two staff suggestions. (MEDSEARCH, having been registered by a firm that had produced other front end software, was not eligible.) The new name, GRATEFUL MED, produced mixed reactions from the day it was selected, but it has proven to be a happy choice, provoking much interest and easy recall. The Beta test materials were quickly modified to reflect the new name.

The library/clinic system test sites were given the option of observing twenty-thirty uses of the software or leaving a self-administered questionnaire for anyone using the system to conduct a search, or a combination of both options. The site coordinators were also requested to provide a one or two page summary of their impressions of the system and its use at their institution. The individual end-user volunteers were asked to either complete the questionnaire or submit written comments summarizing their experience. Free access was provided for all searching and a "help" telephone number provided.

Except for the demographic information, the questionnaire used in Phase I contained open-ended questions that were completed by the observer. To increase the ease with which the self-administered questionnaire could be completed, it was restructured and reformatted to contain mostly multiple choice questions; open-ended questions (comments and suggestions) were optional unnumbered items on the last page of the four-page questionnaire. The draft documentation and copies of the questionnaire were sent out with the software on a floppy disk on December 13, 1985.

Beta Test Results

By February 1986, reports of over 600 documented uses of GRATEFUL MED had been received from ten of the Beta test sites. Coincidentally with the development of the front end software, the National Library of Medicine was celebrating its Sesquicentennial Year. The first major event was a symposium for 100 medical writers and journalists scheduled for February 5, 1986. One of the goals of the symposium was to acquaint the attendees with the services of the NLM that might assist them in their writing, and it was suggested that a Beta test version be distributed at the symposium. External time pressures were now a factor. First, if major problems had been discovered by the Beta testers, it was important to reverse the decision to distribute at the Science Writers' Symposium; second, if distributed as planned to the writers and mentioned by them in print as forthcoming, the release of Version 1 should follow closely.

The Office of Planning & Evaluation undertook the task of data analysis, both from the questionnaires and the site coordinators' summary reports. From the standpoint of stability and robustness, GRATEFUL MED stood the test. From the standpoint of enhancement, the Beta testers had provided even more data than originally expected.

There was a wealth of valuable suggested improvements that would enhance future versions; more importantly, however, a consistent pattern of frequently requested changes clearly emerged. Up until now, the changes and enhancements to the evolving front end had been suggested primarily by the contractor and NLM staff—chiefly the small number of NLM staff involved in acceptance testing. This was the first set of recommended changes coming directly from a large group of potential users, and the process of reconciling the design philosophy and user expectations began in earnest. Some of the requested changes fit nicely with the team's goals—some did not. It soon became apparent that the front end could not serve all populations equally well. The librarians' requests reflected their in-depth knowledge of searching. The end users' requests reflected their lack of knowledge. It clearly would be impossible to "be all things to all people," and imprudent to try.

It also became apparent that all portions of the design philosophy had to be flexible. For example, the principle of uploading the search and downloading the retrieval, and keeping the intellectual processes of the user tied to the microcomputer, not the NLM mainframe, was questioned by one library site in particular. The librarian wanted the option to interrupt the search during the uploading or downloading for the purpose of modification, an act that required knowledge of and experience with the NLM command language. A direct logon to the

NLM computer was allowed through the initial menu, and it was assumed the librarian users who might be making use of the front end as an auto-dialer would elect to use that option. Consideration had not been given to the idea that librarians might want to take advantage of the user-assisted form screen with the option to later modify. Although it was contradictory to the design principle, this capability was added. It was felt that the risk to the end user was small—the occasional end user who might accidently invoke this feature would receive a message including instructions on how to send the "stop command." It was also possible that an ambitious end user might absorb enough from observing the uploading and downloading actually to make use of this real-time modification option. This fit with the design goal of providing the user the ability to "grow" with the system if he or she so desired.

Lesson #5: Even the most sacred design principle can be flexible.

Another example that compromised a point of philosophy involved the display, consideration, and printing of each citation in a retrieved set. The algorithm in GRATEFUL MED that was to guide the end user to appropriate MeSH terms for an expanded search on the same topic was not nearly as appealing to the end user in a hurry as it was to the designers. The student testers, especially, wanted to "print and run" rather than give serious consideration to each citation and print only those deemed relevant to the search question. (One physician suggested that it was not so much the time, but rather that the students did not yet have the experience to know what was relevant.) Of those who were willing to go through the retrieval one by one for printing, many did not want to see or print the MeSH terms. Although a goal of the design philosophy was, and is, to provide MeSH assistance, options were provided to skip viewing and printing MeSH terms, and indeed, to skip the review process altogether. This last change was implemented in the only way possible at such a late date—as a choice in the relevancy question. Because the screen space is limited, the instruction is terse, and perhaps not the optimum implementation.

Lesson #6: Regardless of how much one wants to be responsive to users' expectations, one must weigh carefully whether it is worth making a last-minute change; it might be better to save the change for a future version and implement it differently.

Eight recommended changes warranted serious consideration prior to release of Version 1. The contractor was able to incorporate six of them very quickly. One was already planned for Version 2 (providing MeSH headings during the construction of the search query—a major undertaking that added a second floppy disk to the software). The last recommendation was to put portions of the help screen for zero retrieval (newly added with Phase I testing) directly into the "zero hits" message received for such a search. That is, do not require the user to ask for the help screen, but provide it without his asking. Since the message would get stale with enforced repetition, this suggestion was not implemented, and a better solution to this problem is still being sought.

Three other changes were added in the category of "niceties" rather than "imperatives." It was necessary to issue a sheet titled NEW FEATURES OF GRATEFUL MED, VERSION 1.0 with the *User's Guide,* which had already been sent for printing. This was a problem, as several users did not realize the significance or importance of this two-page addendum. In retrospect, a better approach would have been to print it on bright colored paper and title it BIG, NEW, IMPORTANT FEATURES THAT ARE NOT IN THE USER'S GUIDE!!!

Lesson #7: Do not assume that the users will read every piece of documentation included in the packet. If it is important, make it appear so.

GRATEFUL MED, Version 1, was officially available for order from the National Technical Information Service in March 1986 for $29.95. It had taken slightly more than a year to bring the front end from prototype to release.

Feedback from Library Users

An *NLM Fact Sheet* with the header "GRATEFUL MED: A New Way to Search MEDLINE" was produced and widely distributed. The secondary header, "System Intended for the Individual User," indicated the proposed user population. This proposed user was originally visualized as the individual physician or other health care professional, at home, at night, with his or her microcomputer, doing searches for the patients seen that day. The clinicians among the Phase I Beta testers had reinforced this thinking by their comments. It was, therefore, a happy surprise when a number of medical libraries bought copies and integrated end user searching into the public services/reference setting, with the cost of searching paid by the library budget.

End user instruction for GRATEFUL MED was a natural outgrowth at some of these institutions, and feedback from the librarians and health professionals who have written abbreviated instructions and "brief guides" has been invaluable. Data from the Phase II Beta test showed that, generally speaking, not much attention is paid to written documentation. Material was added to the online help screens where the need for additional instruction was indicated, expanding the Version 1 *User's Guide* only slightly from the one used in Beta testing. The instructional materials developed by library users served to show what some users thought was needed in addition. Better documentation was needed for those who would use it; thus Chris Olson & Associates, an Annapolis, Maryland firm experienced in the production of end user materials, was hired to create a logo and produce both a brochure and a "slick software manual."

Lesson #8: Accept the fact that end user documentation is difficult to write and rarely read. Then try to produce the best possible for those who do read or refer to it, making use of instructions written in the "field" by actual persons using the software.

Feedback from Individual Users

About 25 percent of the requests for new access codes to the NLM mainframe in the first few months of availability of the front end were from individual end users who had purchased GRATEFUL MED (this had grown to 36 percent by August, 1986). Suggestions, some in lengthy letters, were received from a number of end users; they were added to the already growing list of possible enhancements. It became obvious quickly that there were some sophisticated, computer literate end users who wanted the front end to stretch and grow. There were also näive users who had no real interest in learning more than was needed to do a simple search in Version 1. Both groups were important to satisfy.

Lesson #9: If at all possible, do not force users to learn new or changed features in order to use the software the way they first used it. But make the new and changed features appealing so they will want to try them.

VERSION 1 TO VERSION 2

As is often the case, work began on Version 2 before the release of Version 1. Online Information International, Inc., in the continuing role as contractor to NLM, delivered a test copy of Version 2 to NLM three months after the release of Version 1. The two major thrusts of Version 2 were the addition of a permuted MeSH list from which search terms could be selected and the capability to download new versions and/or features from a commercial mainframe (the Source). Both were achieved, but ironically, having the first precludes much of the use of the second.

The ability to select from MeSH had been requested by almost everyone who used the software who had previously used either the printed *Index Medicus* or the NLM's command language online search system. NLM's thesaurus for both indexing and cataloging, MeSH is a highly developed, hierarchical list of over 14,000 terms. These are further divided into major descriptors (appear in *Index Medicus*), minor descriptors (do not appear in *Index Medicus*), and entry terms (cross references to major descriptors that appear in the printed MeSH, but not in *Index Medicus*). The goal was to fit as many of these terms, in a permuted fashion, onto one double-density, double-sided floppy diskette—362,496 characters. Even with compression techniques, it was impossible to include the entire range of the terminology.

To now deviate from using MeSH in its complete format was disconcerting to some staff, but it had to be done. Other than agreement that major descriptors could not be eliminated, there was diverse opinion on how to handle this problem. Alternatives for omitting all or portions of the minor descriptors and cross references were presented and, after discussion, the team opted to exclude the chemical terms from the minor descriptor and cross reference lists. This allowed for the inclusion of minor descriptors and cross references from the remainder of MeSH. The decision was based largely on the needs of the clinician end user community, for whom the extra chemical terms were perceived to be of less potential use than other categories of terms (for example, the disease terms).

Lesson #10: Remember the population to be served, and make design decisions based on best serving that population.

The principle of downloading new modules or versions, which was implemented in Version 2, is still in favor. MeSH, however, is also included on a second diskette. As a result, new versions have been released by mail in conjuction with the annual update of the MeSH vocabulary and the production of a new manual, since it quickly became apparent that it was counterproductive to use the download feature for the dissemination of two full diskettes. The time (to the end user) and the cost (to NLM) are prohibitive. A much better use is for updating or correcting a smaller program module. This has been done and the expectation is that the feature will be used more in the future for this purpose.

Lesson #11: When two major, sweeping enhancements are being tested, it is important to test both simultaneously.

Altogether, nineteen of the twenty-eight suggestions changes on an ever-growing list of potential enhancements/changes were implemented.

More Beta Testing

There was no shortage of volunteers for Beta testing Version 2. Goals for this test were to discover problems in the software and assess the usefulness of the new features. It was decided to retain a few of the Version 1 testers for continuity and enlist some entirely new ones. While testing Version 1, it was learned that it was important to include both institutional and individual users in a Beta test. Individuals can provide excellent insight and well thought-out suggestions for enhancement, but their overall volume of searching is not sufficient for uncovering "bugs." Institutional testers can provide both suggestions and volume use, both dependent, of course, on the individual coordinating the test.

Lesson #12: Include a variety of users in Beta test groups in order to obtain an adequate amount of testing and users who will do insightful searches.

The final constitution of the Beta test group included the following institutional sites as repeats from the Version 1 group: McMaster University, and one of the Regional Medical Libraries, UCLA Center for Health Sciences. New institutions included: the Uniformed Services University of the Health Sciences; Cedars-Sinai Medical Center, Los Angeles; Indiana University School of Medicine; Catholic University School of Library & Information Science; and Ohio State University, Microcomputer Laboratory. The Beta test group also included individual users, some of whom shared the software with others at their workplaces. They received Version 2 in late October, 1986 and tested through December.

In-house staff from MMS had spent about thirty hours in organized testing before Version 2 was distributed, and only one additional "bug" was uncovered by the Beta testers. There were three other important consequences of this Beta test. First, it was discovered that "IBM-compatible" did not always hold true; a number of problems with IBM clone machines were discovered by the testers. Secondly, the much-labored-over decision involving the makeup of MeSH in GRATEFUL MED was confirmed as correct. That alone was worth the test. And third, at the urging of some testers and Version 1 users, two features were added, even though the documentation had already been sent to the printers. One feature was at the request of librarians, and the other, for the health professional users.

The first change was to provide a way to suppress the downloading of MeSH headings (the librarians' request) to reduce the cost of searching. Experiments revealed that the total cost of a GRATEFUL MED search could be reduced 20-50 percent by the omission of MeSH from the downloaded retrieval. This was due to the charging algorithm used at NLM. Suppression capability could be provided for librarians subsidizing end user searching in libraries, but how to do so posed a problem. There was concern that, given the opportunity by a direct question, novice users might elect to "skip" MeSH terms without understanding what they were. The decision was made to implement the function in a subtle rather than overt way, resulting in perhaps the most criticized enhancement undertaken to date. Most users do not know the suppression option exists; of those that do, some do not like the implementation.

Lesson #13: If it is necessary to "protect" the users from a capability, further consideration should be made as to the validity of the change as well as methods of implementation.

The second late addition was for the end users: a way to write citations to a named PC disk file, either new or previously created. This was much appreciated, easily understood, and popular with end users and librarians alike.

Lesson #14: Make a last-minute change only if it is well thought out and intuitively understood without documentation.

Institutionalizing the Product

One of the normal evolutionary changes that accompanies any effort of this type began during Version 2: the institutionalization of the product. What started as a few people in a new venture became a structured group with explicit procedures. This evolved by three steps over a six-month period.

The Office of Computer and Communications Systems (OCCS) had provided management and funding support since the beginning of the project. The Director of OCCS had personally coordinated the development. The first step in building a structured group was in April 1986, when the Director phased himself out and appointed Philip Nielsen as Project Manager and Chair of the Working Group for GRATEFUL MED.

In May, the Associate Directors from the three program areas at NLM involved in user services instituted a procedure to prioritize the "wish list" for enhancements. The Associate Directors from Library Operations and Specialized Information Services join the Director of OCCS in regular meetings to consider the more formally named "GRATEFUL MED Priority List." Convening this group provides an opportunity for management to weigh the costs and benefits of the proposed changes and assign priorities before the list is submitted to the Director of NLM.

The third step in the institutionalization came in October 1986, when an official GRATEFUL MED Working Group was established, composed of representatives of several Library areas, and the contractor, whose suggestion it had been to form a Working Group.

This Working Group since that time meets biweekly, often for three hours or more. In terms of staff time, it is an expensive venture. More diverse perspectives lead to lengthier decision-making processes. Overall, however, these meetings are a workable method for maintaining a team approach.

Lesson #15: To encourage staff participation in and acceptance of a new product, involve all program areas in the institution who have a vested interest in the outcome, both at management and working levels.

Integration into Operations

GRATEFUL MED was further institutionalized when it was installed in the NLM Reading Room for online access to the card catalog (CATLINE) as well as for MEDLINE (journal) searches. Since NLM library patrons, many of whom are repeat users, were accustomed to the natural language CITE system for CATLINE, a gradual transition was provided to GRATEFUL MED. The outdated hardware used for CITE was impossible to replace and increasingly difficult to have repaired. One microcomputer was installed in the Reading Room and user aids were developed for both the software and the PC. A second PC was installed several months later; in November 1987, all of the CITE terminals were replaced with GRATEFUL MED on PCs.

Staff of the MEDLARS Management Section were also heavily affected by the software, particularly by increases in the number of inquiries for information, paperwork involved in adding new users to the system, and service calls for assistance. There has been a marked increase in all of these activities, especially immediately following a new release. In addition, with Version 2, MMS was named directly responsible for in-house testing, a time-consuming task.

Reference and MEDLARS Management are two groups in the Library who observe or talk with end users directly about the software on a regular basis. Both have representatives on the Working Group, facilitating the transfer of direct observation and experience to the evolution of the software.

Lesson #16: Provide a means for staff who regularly interact with the users to routinely report feedback.

One huge facet of software production is documentation. The original idea of providing only minimal paper documentation and maximum Help screen information was finally modified by users who repeatedly requested a "real manual." The documentation contractor had designed a total software package, including a three-ring *User's Guide* with diskettes in vinyl pockets, all in a slipcase. A subset of the Working Group had worked closely with the contractor to produce the text; layout and formatting were done by the contractor. The project remained on schedule until a change in the regulations at the Department of Commerce, the parent organization of the National Technical Information Service, caused a three-month delay in printing.

Lesson #17: Allow extra time in the schedule for all portions of the production that are not in your complete control.

Version 2 was released in March 1987, thirteen months after Version 1.

VERSION 3

GRATEFUL MED, by now, had become an institutional product with a pace, a cycle, and a momentum of its own. Work on a new version again was begun before the preceding one was mailed to users. Using the now established procedures, the Working Group had drawn up a list of potential enhancements which was reviewed by the Management Group and given to Dr. Lindberg for final review. The highest priorities were: making more databases searchable via GRATEFUL MED, adding a search "edit" capability, and implementing an auto-install feature. In addition, Dr. Lindberg and the contractor worked together to design what came to be known as the GRATEFUL MED Search Engine, a software routine that allows system developers to incorporate the GRATEFUL MED search and retrieval capabilities into their own PC applications. The Search Engine has particular application for researchers from the hospital and academic community working under an NLM contract on the Library's ongoing Unified Medical Language System (UMLS) project.

During the development of this version, database experts from DIRLINE, TOXLINE and TOXLIT, CANCERLIT, CHEMLINE, HEALTH PLANNING & ADMINISTRATION and AVLINE were added to the Working Group to contribute and test the implementation of their particular files. Having so many extra hands led to a mistake: when it came time to Beta test, in-house testing was thought to be sufficient. Although sixty hours of formalized testing were done, there have been errors discovered in Version 3, which in retrospect would more likely have been found in outside Beta testing.

This was an especially hard-learned lesson because the Version 2 Beta test had revealed some equipment incompatibility problems. A lesson from the past had been ignored.

Lesson #18: Don't assume that Beta testing can be done in-house. A wide variety of users and equipment and an extremely high volume of searches are imperative.

The only delay in meeting a mid-December distribution date (to coincide with the annual update of MeSH terms) again involved printing. Sixty percent of the *User's Guide* pages contained new material, so it was decided to reprint the entire manual rather than require the user to replace pages. The printing process caused only a three-week delay this time; Version 3 was distributed in early January 1988, ten months after Version 2.

By now it was clear that the Version releases are inexorably related to NLM's MEDLARS annual releases.

Improvement by Analysis

Up to now, improvements have been based primarily on user feedback, within the constraints of the design philosophy. That method of operation will be maintained; in addition, a more systematic assessment of the effect of a change will be made before implementing it.

To do this, a PC software log program has been written that collects data on how the user interacts with the software and the results of the search, but does not include data on the actual search content. Data on searches done in NLM's Reading Room with the current version will be collected and compared with data collected from a test version that contains a proposed change or enhancement . If the change does not produce the contemplated result, e.g., ease of use or improved retrieval, its implementation will be seriously questioned. Data collected for 5,000 searches in the Reading Room show that only 6 percent of the users looked at the Help screens; 20 percent used MeSH terms in their subject searches; and 40 percent of the searches retrieved nothing. This problem of zero retrieval will be the first to be approached with the new test stations. Nothing in the literature reports what fraction of the time searches generally retrieve nothing, but 40 percent is at least 25 percent too high. Significantly, this represents a pattern totally different from that seen for years with searches mediated by a medical librarian. These tend to have the opposite fault: that is, initial retrieval of far too many citations. There are various ideas for solving the new problem; each will be tried and data collected on the resultant change.

Both full function and test versions will be available side-by-side in the NLM Reading Room and at the National Institutes of Health Library, which is heavily utilized by clinical researchers. Consideration will be given to adding other test locations as this method evolves.

At this point in the development, the goal is to make the software "smarter" and more responsive to the end user's problems with searching. Plans are underway to construct a "hook" to an expert searcher program residing outside GRATEFUL MED. The results of the user's search attempts will be available to the expert program for analysis leading to suggestions for improving the search, particularly in the use of vocabulary.

Other Evaluation Strategies

NLM has three evaluation projects underway that will undoubtedly influence the future development of GRATEFUL MED. A nationwide survey of nearly 3,000 health professionals who search the MEDLINE database, either by command language or using a front end such as GRATEFUL MED, is still in the data analysis phase; but preliminary results have underscored the importance of MEDLINE searching for patient care—69 percent indicated patient care as a primary purpose of their searching. This reinforces the goal of designing a system that best serves the information needs of the clinician.

A more elaborate study to collect information from health professionals about their use of MEDLINE using the Critical Incident Technique (CIT), an evaluative methodology that systematically collects and analyzes reports of users' actual behavior, is just beginning. In simplest terms, the CIT is used to determine critical requirements that have been demonstrated to make the difference between success and failure in carrying out an important part of a task. The goal is to understand and document how MEDLINE information is used, especially in patient care, and with what effect(s). That is, does the use of MEDLINE make a difference? The study results will be applied to improving the design of both the command language system and GRATEFUL MED.

A third project now being planned is to make use of a laboratory facility where users' behavior can be systematically observed under controlled conditions. Tasks will be designed that will provide information on how test subjects use the software and documentation while observed by "usability lab" staff. Data will be collected by recording a user's behavior/comments in a real-time computer log and by videotaping the interaction for later analysis. A GRATEFUL MED tutorial now under development is a likely candidate for this type of evaluation.

CONCLUSION

Rapid changes in micro and large computer technology and telecommunications capabilities, as well as the increasing level of computer sophistication in the health care professional community, promise ever greater levels of utilization of computer-based information resources for the nation's health care. Working on the design, development, and evaluation of GRATEFUL MED has been exciting and satisfying. But it is just a beginning.

[Authors' note: As of October 1990, over 29,000 copies of GRATEFUL MED have been distributed.]

ACKNOWLEDGMENTS

The authors wish to acknowledge the following persons who so generously contributed to the successful testing of GRATEFUL MED.

The following site coordinators participated in Phase II of the testing: Arthur Downing, New York Academy of Medicine; Frieda Weise, University of Maryland; Phyllis Self, University of Illinois at Chicago; Marie Reidelbach, University of Nebraska; Tricia McKeown, University of Texas at Dallas; Kay Denfield, University of Washington; Julie Kwan, UCLA; R. Brian Haynes, M.D., Ph.D., Ann McKibbon, and Lynda Baker, McMaster University; Joyce Mitchell, Ph.D., University of Missouri; Eleanor Goodchild, University of Pennsylvania; and Mary Horres, University of California, San Diego. Thanks are also due the following individuals who volunteered as end users: Randolph Miller, M.D.; Morris Collen, M.D.; Alfred Fishman, M.D.; Douglas Brutlag, Ph.D.; and Richard Friedman, M.D. In addition, Brian Haynes and Ann McKibbon of McMaster University and Julie Kwan from UCLA participated in the final Beta test group. New participants in this group were: Commander Joseph Henderson, M.D., Uniformed Services University of the Health Sciences; Phyllis Soben, Cedars-Sinai Medical Center, Los Angeles; Frances Brahmi, Indiana University; Barbara Rapp, Ph.D., Catholic University; and Gordon Black, Ohio State University. The Beta test group also included individual users Carolyn McHale, Donald Hawkins, and Andrew Dean, M.D.

REFERENCES

Abrutyn, E. (1986). Software 1.0. *Annuals of Internal Medicine 105*(August), 321.

Bonham, M. D., & Nelson, L. L. (1988, January). An evaluation of four end-user systems for searching MEDLINE. *Bulletin of the Medical Librarian Association, 76*, 22-31.

Bryant, J. (1987). GRATEFUL MED: A tool for searching the biomedical literature. *BioTechniques, 5*, 468-70, 472, 474.

Haynes, R. B., & McKibbon, K. A. (1987). GRATEFUL MED. *M. D. Computing, 4*(September/October), 47-49, 57.

National Library of Medicine. (1987). *GRATEFUL MED User's Guide.* Bethesda, MD: NLM.

Schell, C. L., & Rathe, R. J. (1988). Computer searching made easy. *Postgraduate Medicine, 83*(January), 325, 328, 331-332.

Snow, B.; Corbett, A. L.; & Brahmi, F. A. (1986). GRATEFUL MED: NLM's front end software. *Database, 9*(December), 94-96.

Watson, L. (1987). GRATEFUL MED version 2.0: An overview for searchers. *Medical Reference Services Quarterly, 6*(Summer), 1-16.

Wigton, R. S. (1987). Literature searches: On-line data bases get easier to use. *Computer News for Physicians, 5*(October), C19-C21.

JOHN J. REGAZZI

President and CEO
Engineering Information, Inc.
New York, New York

The Uses of CD-ROM and Other Information Delivery Systems for Libraries: A Publisher's View

INTRODUCTION

The delivery of electronic information to libraries is increasing significantly in both volume and forms of delivery. These forms of delivery now include online searching, local access systems, and CD-ROM, to name only a few. CD-ROM is a technology, however, that is growing the fastest and has recently generated the most excitement in the library and publishing communities.

This excitement is balanced by some concerns. Such concerns are diverse. This paper, however, addresses one concern in particular. That is, how will CD-ROM and other modern information handling technologies affect electronic publishing programs in general, and therefore by extension, libraries and publishers as well? Although the views represented here are of a publisher of CD-ROM systems, these views are not intended to suggest that these technologies are suitable in all publishing environments, or for all libraries. The comments in this article are simply intended to describe significant developments and trends in electronic publishing.

This article consists of three parts: case history, historical perspective, and speculation. To start, a brief case history of the H. W. Wilson Company's electronic publishing program is in order.

CASE HISTORY: THE H. W. WILSON COMPANY

Until recently, H. W. Wilson produced its publications using linotype machines and the interfiling of the "lead lines of type." It took the company a long while to automate, but now all the lead is out of its plants and fully automated methods of acquiring and producing information are in use.

In making this transfer there was a need to develop a comprehensive system of producing, editing, managing, and electronically typesetting this information (see Figure 1). But such a system is only the first step in the use of today's technology, and certainly does not itself constitute a viable electronic publishing effort. As is evident from Figure 1, this automated system provides the basis of not only producing and typesetting information electronically, but also includes an electronic dissemination program. The publishing program includes the use of online systems as well as personal computers and CD-ROM technology. This new, emerging role of the publisher or producer of the information as a more active agent in the electronic dissemination process will be dealt with in this paper.

Figure 1
The Wilson System

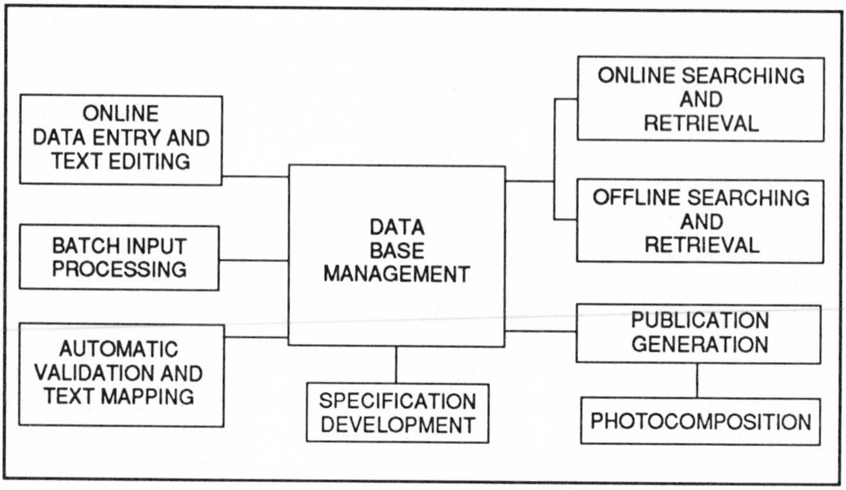

ONLINE DATA ENTRY AND TEXT EDITING
BATCH INPUT PROCESSING
AUTOMATIC VALIDATION AND TEXT MAPPING
DATA BASE MANAGEMENT
SPECIFICATION DEVELOPMENT
ONLINE SEARCHING AND RETRIEVAL
OFFLINE SEARCHING AND RETRIEVAL
PUBLICATION GENERATION
PHOTOCOMPOSITION

HISTORICAL PERSPECTIVE: PUBLISHERS—INNOVATION THROUGH CHANGE

Publishers are facing significant changes in the publishing and information industry. Publishers of abstracting and indexing information, for example, are seeing exponential growth in the primary literature, i.e., more information is being produced and therefore more information needs to be organized. They are also experiencing more competition, i.e., more information goods and services are in the marketplace. Perhaps most significantly, they are experiencing greater and greater demands from their clients in order to meet the changing needs of information seekers. These demands tend to be for more accurate, current, diverse, sophisticated, and especially, economical information goods.

Most publishers who have begun to meet these changes successfully have done so through the utilization of new and more sophisticated technology. This trend will not change. The direction of these technological developments is critical to publishers and libraries alike. The use of more technology in publishing will not only affect the types of information products developed and made available, but also and perhaps more importantly, will fundamentally change the delivery systems for information goods and services.

Technology will effect this change mainly for two reasons: One is that the innovations in information technologies of the past two decades have radically reduced the time and cost of processing information. Second is that, by reducing particularly the costs of processing, information technology will lead to an overall shift toward proportionately more use of markets rather than the highly centralized and hierarchical delivery of information goods and services which presently exists.

There are basically two methods of organizing an electronic publishing effort: through markets or through a central source.

Markets coordinate the flow of information goods and services through the coordination of different individuals and organizations. Market forces will tend to regulate this flow through the traditional economic gauge of supply and demand. In a market environment, information products may tend to be easily linked to other products to meet complex market demands more efficiently. Thus, the consumers of information products may acquire specific information goods from a variety of sources, and make choices based upon the best combination of product attributes to meet individual needs.

A central source, on the other hand, tends to provide information goods through a single channel or chain of distribution. Although in many cases two legal firms are involved (i.e., a producer and a vendor), the producer of the information licenses a vendor to disseminate the information, and the vendor takes primary, if not exclusive, responsibility for those information goods. Managerial decisions, not necessarily market forces, determine the design, price, and availability of the product. Buyers generally do not select a supplier from a pool of vendors, but rather from one or a few select vendors.

There are, of course, variants of these two basic approaches, but most electronic publishing programs today tend to use the "centralized" approach. Electronic publishing efforts in the future, however, may be able to make a much more market-oriented approach through the use of the technological developments which have taken and are taking place today.

Online Search Services—A Characterization

The key characteristics of online search systems in the period 1970-85 are as follows:

— Highly centralized database services usually operated by a single organization, at a single data center, generally running on a single mainframe computer. All of the data and software resides at the data center location.

— Data centers usually supported by standard "unintelligent" terminals through a limited number of telecommunication carriers (sometimes owned by the vendor). These networks support character transmission rates of 30 to 120 characters per second.

— Scope, searchability, and design of the data base and information sources generally decided by negotiation between publisher and vendor. Price, currency, and frequency of update also tend to be managerial decisions of the vendor which may have been influenced through the negotiation process.

TECHNOLOGY AND ORGANIZATIONAL FORM

The choice of organization is primarily an economic decision which depends in large measure on four key factors: production costs, coordination costs, product complexity, and asset or database specificity.

Production costs are intended to include all of the development, editing, manufacturing, distribution, packaging, marketing, handling, fulfillment, and other costs associated with the delivery of an information good.

Coordination costs, on the other hand, take into account all of the transaction costs in developing the information good. These costs include, for example, the costs of gathering information about suppliers, the market, and so forth; the costs of negotiating contracts; the costs associated with coordinating the work of people and machines operating in different companies; and all costs associated with the transfer of goods between producer and supplier.

Interrelationship of Production and Coordination Costs

The interrelationship of production and coordination costs, as indicated in Figure 2, affects the choice of certain organizational forms and favors certain organizational forms at different levels of expenditures. For example, production costs for the publisher are higher if the publisher produces the product alone, but the more suppliers and distributors a publisher needs to utilize, the higher the coordination costs. This may explain, in part, why publishers have heretofore tended to offer their electronic products through a highly centralized method.

Figure 2

Relative Costs of Markets and Integrated Efforts

Organizational Form	Production Costs	Coordination Costs
Markets	Low	High
Integrated	High	Low

In the 1970s, and certainly by 1980, most publishers had already absorbed the high automation and production costs associated with the development of information goods for their printed products. Electronic typesetting was and is widespread in the industry, but electronic publishing was considered in many cases a byproduct of the print production process. Production costs had already been assigned to and absorbed by the printed product. The costs of database design, storage, online access, testing, and marketing for an external searching system, however, were formidable and outside of the traditional strengths of publishers. A substantial investment of coordination costs was required by both a publisher and a vendor to make these databases available. Thus, many publishers opted for one service only. In the early days of online searching, it was not uncommon to find exclusive database licenses between publishers and vendors.

Other Factors Affecting the Choice of Organization

Other factors which further influence the form of organization are asset or database specificity, and product complexity.

Database specificity is intended to describe the uniqueness of the information good, and thus provides an indication of how difficult it could be to process a publisher's information resource or database by a third party. Some factors which might make this difficult are, for example, a highly complex data structure, unique hardware requirements, difficulty in transporting the data from one data center to another, or the inclusion of uncommon character representations such as typesetting codes or other special characters.

Product complexity, on the other hand, provides an indication of the amount of information needed to specify the attributes, functions, and features of the service to be provided. Obviously, the more access points, search rules, and other functions required in an information service, the more complex the product.

Relationship of Asset Complexity and Product Coordination

As one looks at the relationship of these factors, one can find how these factors may have influenced the choice of organizational form (see Figure 3).

When both the information resources (assets) and the products which are to be derived from those resources are highly specific and highly complex, the centralized approach is more desirable, since the publisher or its agent is best able individually to deal with these complexities efficiently and in a cost-effective way. As more suppliers are introduced, the costs associated with the peculiarities of the data and the product are replicated.

When data can easily be transferred to suppliers and when these suppliers can easily manipulate this data in a variety of ways, a market approach is easily effected, and the high costs of data conversion and processing are reduced.

To reference the online industry in the 1970s and 1980s, the production systems which publishers used to typeset their printed products tended to be highly specific to an individual publisher. As a

Figure 3
Product Attributes Affect Types of Organizations

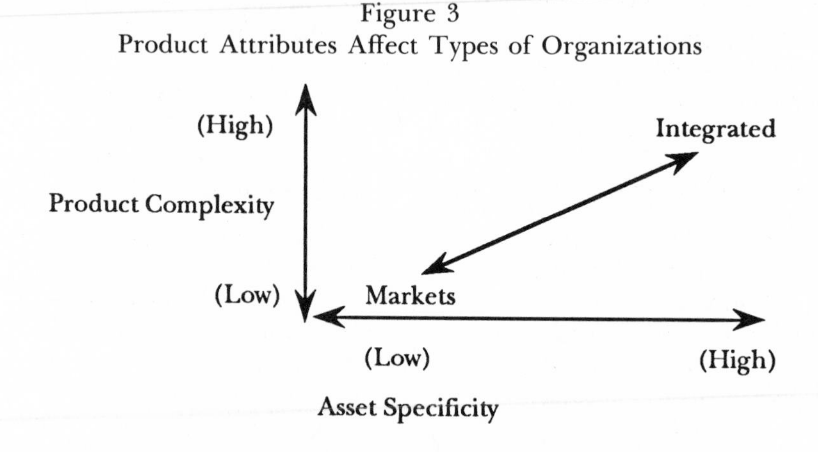

result, publishers had little capability to provide their information resources in a standard or flexible form. Second, database products tended to be as varied as the number of databases offered. The outputs of individual production systems, data structure and form, and the access and searchability of the data itself were all very different for each database.

Relatively low production costs but high coordination costs, in conjunction with the very specific information resources and complex data structures, naturally and logically changed to a highly centralized approach in the period under consideration.

New Directions

What is different today and for the future? Information technology will continue to provide an impetus for change in the publishing industry. CD-ROM or optical disc is a significant technology for publishing, but it is by no means the only significant information handling technology which needs to be discussed for the future. In fact, CD-ROM technology, in order to be successful, by its very nature must be used in conjunction with other technologies in order to provide effective new information products.

The discussion of the future directives and uses of CD-ROM has been divided into three areas: Technology, Systems, and Economics. Each of these is changing in ways that will significantly affect the use of optical disc technology specifically, and the development of information systems and services generally.

Technology

In the area of technology, three key technologies are considered for their effects on electronic publishing: Computational Capabilities, Storage Media, and Communication Capabilities. To illustrate the advances in each area since 1970, Figure 4 describes a cost comparison for 1970 and 1987 in the per unit cost of these technologies for each of these years.

What are the implications of these dramatic cost reductions? First, the same technology that launched the online industry is now available at a fraction of the cost. This means that the technology for providing information goods and services is available to many more organizations. More publishers and other organizations are able to afford the basic tools of electronic publishing programs. Second, technology keeps developing. For example, PCs continue to increase on the price/performance curve, and this beyond even the comparisons offered in Figure 4. The IBM Personal System/2, Model 30 introduced recently

Figure 4

Price/Performance Comparisons: 1970-1987

	1970	1987
Computational Technology (CPU instructions per second)	$3.00	$.15
Storage Technology (per million characters)		
Magnetic Disks	$2,500	$19
CD-ROM (Master)		$ 6
Telecommunications Technology (per million characters transmitted)		
Dial Access	$1,000	$200
Leased Line		$ 10

will be replaced by a machine which has nearly 100 percent more throughput and is projected at a cost reduction of 25 percent. It is also an increase by 200-300 percent of capacity from the PC/XT and AT machines with comparable cost savings of nearly 50 percent. Another example of developing technology is the erasable optical disc, which provides high density optical storage on a reusable basis. A third technological development is the expansion of communication networks. For example, CONNET was announced a few years ago. It is a cooperative project between Southern New England Bell and Telenet where, with the use of X75 software, one can connect to a Telenet mode from anywhere in the state of Connecticut for little more than a local telephone call.

System Direction

As these technologies advance, systems design continues to become more powerful and sophisticated. This development enables these systems to provide more information, faster and far less expensively than present information delivery systems.

Systems are being developed around networks which utilize intelligent personal computers in conjunction with mainframes, creating a socalled micro/mainframe connection or a local area network (LAN). In such a networking environment, local information can be processed easily and inexpensively without significant telecommunications and computer costs. Such designs are capable of accessing a larger centralized data resource on the system when needed and on a far more economical basis. Many of the front end search systems, post-processing software, and so-called hybrid CD-ROM systems have effectively utilized this principle.

Within this system network orientation, there is also a concerted effort among major computer manufacturers and suppliers to permit computers to transfer data easily across systems and across manufacturers' models. The computer industry has begun discussion on a common computer protocol interface which is being supported by the major manufacturers, and which has as its goal a standard method of transferring data for all computers. The gateway systems and the intelligent networks which have already begun to emerge will greatly benefit from such developments. Publishers, moreover, may find it increasingly feasible to maintain their information resources in their own systems, while also increasing dissemination efforts through a variety of networks and gateway relationships.

Finally, systems are becoming richer in the types of information that can be provided. With the increased capability of optical discs and the proper standards to utilize them fully, multimedia systems are emerging. These are information services which provide not only textual and numeric information but also video, graphic, and audio information. Systems are increasingly being designed to simulate different views of the data. In other systems, speech input and touch screens are being introduced to reduce or obviate the need for keying information and/or the knowledge of specific system commands and syntax.

Economic Directives

Perhaps the most significant factors influencing the changes that will occur to information systems have to do with the market and the needs of information seekers. These factors are primarily economic.

Electronic access to information is increasingly being demanded by audiences different from the traditional online users. These new and potential users, however, appear to be far more price-sensitive to information goods than the traditional online consumers.

What complicates the development of new information delivery systems further is the ability of traditional institutional clients to finance the consumption of these services. Although the cost of information goods may be decreasing and the value of the information services increasing, public, private, and corporate monies are being stretched further and further for information acquisition. This stretching appears to be particularly true for the public and voluntary sectors, especially public, school, and university libraries.

In short, the need for access to information is widespread, but the single greatest obstacle to the use of electronic publishing systems by new markets is the high cost of those services. Publishers will continue to be challenged by this economic directive in how they produce information and new information products. The most significant costs to a publisher are still the intellectual costs associated with the acquisition, organization, and editing of the information. Duplication of effort by and among publishers will also be costly. As publishers seek to develop new information products in order to meet new client needs, they will seek new partnerships with other publishers. These partnerships will undoubtedly be difficult to forge, but the benefits to publishers and libraries alike will be substantial.

Optical Discs—Conflict/Confluence

Optical discs, and the CD-ROM technology specifically, must be considered within the broad context of these economic and system trends. A major question for publishers is how to integrate these technologies into their existing publishing programs, both printed products and existing online database services. Similarly, how should libraries view and use these new CD-ROM-based services? In effect, will CD-ROM replace or complement existing information delivery systems, or will it will a short-lived technology?

The major advantages to the CD-ROM technology are three: first, CD-ROM provides publishers the ability to disseminate their services in new ways; second, this new method of distribution provides new value through improved methods of access while providing greater use of data already created; and third, the manufacturing costs associated with CD-ROMs are relatively inexpensive. The advantages to consumers are that more information and more sophisticated systems can be acquired through the use of this technology and that these systems offer a tremendous economy of scale, unlike their online counterparts, and are also far less labor-intensive.

What are the disadvantages or dangers to CD-ROM publishing? To the publisher, these dangers center more around the changing technologies than in the CD-ROM itself. One disadvantage is that publishers must make added investments in system development, training, and support services for these systems. These activities and investments are not traditional strengths of publishers. A second problem is that in a changing technological environment, publishers will constantly be challenged in how to manage the transition of their businesses from older to more advanced technologies.

Although libraries and other consumers of these products will generally benefit from advances in information technology, there are at least two areas of concern: cost and performance. The acquisition cost of these systems is higher than existing services. Unlike online services where one can acquire and purchase only that information which is needed, CD-ROM systems require ownership of the entire database. Although there is a savings in labor to support these new systems, there is also a shift from labor-intensive activities to the capital-intensive activities of acquiring and maintaining these new information technologies.

The issue of system performance also needs to be monitored carefully. The ability to meet the needs of clients in a timely and responsive manner is critical to the success of any system. CD-ROM can only be measured on how well it increases the price/performance ratio of information delivery systems.

CONCLUSION

There is a high demand for change in the publishing industry, and these changes are being driven by technology and economic concerns. As the technology advances, it appears that information goods and services will be delivered in a more market-oriented environment. In this environment, consumers of information will have a much wider variety of options in the consumption of information goods and the characteristics of these goods. Finally, CD-ROM is an important and significant first step in these new evolving technologies, but present CD-ROM systems must be viewed within a broad context of a market orientation to information delivery.

CONTRIBUTORS

WALT CRAWFORD is a Senior Analyst in the Development Division of The Research Libraries Group, Inc. He is the author of nine books on various aspects of library automation, including *MARC for Library Use*; *Patron Access: Issues for Online Catalogs*; *Technical Standards: An Introduction for Librarians*; and *Desktop Publishing for Librarians*. He is a frequent contributor to several publications in the library field and edits the *LITA Newsletter* and *Information Standards Quarterly*.

M. LESLIE EDMONDS is Youth Services Coordinator at the St. Loius Public Library and was previously Assistant Professor at the Graduate School of Library and Information Science at the University of Illinois at Urbana-Champaign. She received her Ph.D. in Curriculum and Instruction from Loyola University and a master's degree in Library Science from the University of Chicago. She received the first Carroll Preston Baber Award with Paula Moore from ALA to support her research of children's use of library catalogs. She is currently developing training for children on how to use and understand online catalogs.

KATHRYN LUTHER HENDERSON, Professor, Graduate School of Library and Information Science, University of Illinois at Urbana-Champaign, has been teaching courses in cataloging, classification, bibliographic organization and technical services for twenty-five years. Recently, with William T Henderson, she has developed a course on the preservation of library materials. Among her publications are *Conserving and Preserving Library Materials* (UIUC, GSLIS, 1981) and *Conserving and Preserving Materials in Nonbook Formats* (GSLIS, UIUC, 1991), both co-edited with William T Henderson.

WILLIAM T HENDERSON is Preservation Librarian and Associate Professor of Library Administration in the University of Illinois at Urbana-Champaign. For more than twenty-five years, he has worked on a system-wide basis with binding and preservation in a large university library. He has incorporated preservation considerations and requirements into the library's binding contracts and procedures, and has been instrumental in developing a local in-house mending and repair facility into a conservation facility. He is a member of the Library's Preservation Committee; has been consultant for other institutions and agencies on the development of preservation methods and programs; and, with Kathryn Luther Henderson, teaches an annual course on Preservation of Library Materials in the Graduate School of Library and Information Science. He is co-editor of two books with Kathryn Luther Henderson:

Conserving and Preserving Library Materials (GSLIS, UIUC, 1981) and *Conserving and Preserving Materials in Nonbook Formats* (GSLIS, UIUC, 1991).

RALPH E. JOHNSON is Assistant Professor in the Department of Computer Science at the University of Illinois at Urbana-Champaign. He received a Ph.D. in Computer Science from Cornell University. His major research interest is object-oriented programming, and he has applied it to a number of areas, including user interface design. He teaches courses in user interface design, object-oriented programming, and other areas of computer science.

W. DAVID PENNIMAN is Libraries and Information Systems Director at AT&T Bell Laboratories. Prior to joining Bell Laboratories in 1984, he served as Vice President for Planning and Research for Online Computer Library Center (OCLC) where he established the Research Department in 1978. He has worked as a research scientist at the International Institute for Applied Systems Analysis in Austria and the Battelle Memorial Institute in Columbus, Ohio, where he was also Associate Manager of the Information Systems Section responsible for the development of the BASIS online retrieval and data management system. He holds a B.S. in Engineering and a Ph.D. in Communication Theory from Ohio State University. His experience in the field of information systems research and development has included involvement in the areas of human factors, human-computer interaction, computer networks, and information systems evaluation.

JOHN J. REGAZZI is President and CEO of Engineering Information, Inc., a not-for-profit information services and publishing company founded in 1884. Previously, he was Vice President of the H. W. Wilson Company responsible for computer services and electronic publishing. He received a Ph.D. in Information Studies from Rutgers University.

RICHARD RUBIN is Assistant Professor at the School of Library Science at Kent State University. He received his Ph.D. in Library and Information Science from the University of Illinois at Urbana-Champaign, and his M.L.S. from Kent State University. He is author of the book, *Human Resource Management: Theory and Practice*, and was formerly Personnel Director of the Akron-Summit County Public Library.

ELLIOT R. SIEGEL is Assistant Director for Planning and Evaluation at the National Library of Medicine. He has been with the

Library since 1976, starting as research scientist in NLM's research and development division and currently director of the Office of Planning and Evaluation, where he is responsible for the Library's long-range planning, evaluation research, and outreach activities. He received a Ph.D. in Communication and an M.A. in psychology from Michigan State University.

MARTIN A. SIEGEL is Associate Professor in the Graduate School of Library and Information Science, University of Illinois at Urbana-Champaign, and Assistant Director of the Computer-Based Education Research Laboratory (CERL) at UIUC. His fields of interest are computer-human interface design, computer-based education, instructional design, and reading comprehension and literacy. Publications include *Understanding Computer-Based Education*, with D. M. Davis (Random House, 1986); and "Redefining a Basic CAI Technique to Teach Reading Comprehension" in *Computers and Reading* (Teachers College Press, 1987). He is currently on leave and working for Authorware, Inc., a Minneapolis-based high technology vendor specializing in next-generation software products and services.

LINDA C. SMITH is Associate Professor in the Graduate School of Library and Information Science at the University of Illinois at Urbana-Champaign. She received her Ph.D. from the School of Information Studies at Syracuse University. Her fields of interest include information retrieval, library automation, and science reference service.

JESSICA R. WEISSMAN is Manager of Computer-Based Training for James Martin Associates. She has been working with computer-based instruction for fifteen years in university, government, and corporate settings. Her Hypercard projects include the Library of Congress' American Memory interactive pictorial database. She has designed and programmed interactive instruction, managed projects, supervised quality assurance, evaluated authoring tools, and designed authoring systems.

ROSE MARIE WOODSMALL is a Program Analyst in the National Library of Medicine's Office of Planning and Evaluation. She has an M.L.S. from Indiana University and has held several positions at NLM involved with the development, management, teaching, and evaluation of the MEDLARS information retrieval system. Her continuing interest is end user searching, especially the integration of user opinion into development.

INDEX